中英日對譯

Questions and Answers about Science in Simple English

圖解科學Q&A

三悅文化

在我們生活周遭，充滿了眾多的『疑問』。其中有許多似懂非懂，好像知其然、卻不知其所以然的疑問存在。

舉例來說：「宇宙的盡頭究竟是什麼？」、「天空為什麼是藍色的？」、「動物為什麼有分雌雄？」、「人類為什麼一定會死亡？」、「行動電話究竟是透過什麼來聯繫？」……諸如此類。

本書將透過科學的角度來解答這些看似普通平常，答案卻非常深奧的疑問。除此之外，用平易近人的文字，加上中學程度的簡易英文來說明，是首次進行的嘗試。

有關宇宙或地球的疑問、包含人類在內的生物疑問、與我們的身體有關的疑問，除此之外，還有與我們生活息息相關的所有疑問等等，共挑選79個項目。

在內容呈現方面，採用「問」與「答」的方式撰寫，在與答案相關的所有解答中，將優先採用代表性的事物與最新的學說。因此，本書裡所解說的答案並非獨一無二，也許會有其他的回答方式或者學說，在此要事先聲明。

本書中每句英、日文皆有標號，各位讀者在閱讀時有無法理解的地方，立刻就可以發現下面有中文的說明。

關於外語學習方面，如果主題與內容不生動的話，是沒有辦法持續閱讀的。也就是說在一面閱讀本書—「科學的疑問」時，不知不覺當中要是能夠增強英、日語的能力，那就是令人再高興不過的事了。

最後，能夠完成本書，要感謝山梨大學的物理學教授松森靖夫先生，還有同校英語教育系的古家貴雄先生的監修。關於英文方面，特別還接受古家先生的同事—英國籍講師的建議。藉著這個機會，在這裡致上最誠摯的感謝。

PREFACE ／前言

There are various questions which we can not understand around us. Some of them are the questions which we think we understand but actually we do not understand and which we think we can answer but actually we can not answer well.

Some examples of them are "What is there at the end of the universe?","Why does the sky look blue?", "Why do animals have two sexes, male and female?", "Why must all human beings die?", "How are cell-phones connected?", and so on.

Such questions, which are simple but we can not answer well, are scientifically clarified in this book. Furthermore, this is the first book to answer the questions both in simple Japanese and in simple English, which even junior high school students may understand.

79 topics about universe, the earth, creatures including human beings, our body, and many things in our daily life are picked up.

This book adopts the questions and answers style. When a question has various believable answers, the typical and latest theory is introduced. Therefore note that the answer introduced in this book is not the only one and some questions have many other answers and theories.

Every Japanese and English sentence is numbered in this book. Therefore when you have difficulty in reading the sentences, you can find their Chinese translation in no time.

You may give up studying foreign language easily if the textbook does not have interesting subjects and contents. From this point of view, it would be more than happy if you can brush up your English and Japanese unconsciously by reading the very interesting science questions and answers in this book.

In closing, in the course of making this book, Professor Matsumori Yasuo of University of Yamanashi, the specialist in science, supervised the Japanese part and Professor Furuya Takao of the university, the specialist in English education, supervised the English part of this book. Professor Furuya's British colleague also helped us with the English part. I would like to express special thanks to them all.

CONTENTS／目錄

Questions about the earth/ 地球の疑問（地球的疑問）

Chapter 3

Questions about creatures ／ 生き物の疑問（生物的疑問）

Questions about human body/ 人間の体の疑問（人體的疑問）

CONTENTS／目錄

Chapter 5

Questions about the things around us/ 身の周りの疑問（生活周遭的疑問）

CONTENTS／目錄

Chapter 1

Questions
about
the universe

第 1 章

宇宙の疑問
宇宙的疑問

Q 1 　When and how the universe was born?

①There have been a lot of unproved theories about how
　　　　　　　　　　　　　尚未證實的理論　　　　　　　　假說
the universe was born. ②The most popular theory of all
宇宙是如何形成的
is the Big Bang theory. ③According to this Big Bang theory, the
　　大霹靂理論　　　　　　　　　根據
universe was originally an extremely small dot, which was red hot
　　　　　　　　　　　一個極微小的點　　　　　　　　　高溫
and had a very high density. ④This theory says that the dot began
　　　　　　　　高密度
to get bigger and bigger very rapidly in time and space
　　逐漸擴張　　　　　　　　　　　　　　在時間與空間
as if it had blown up, and then the universe was born.
宛如發生大爆炸一般
⑤In many different ways, scientists are studying about when the
　　在不同的研究當中
universe was born. ⑥An electromagnetic wave named CMB
宇宙是何時誕生的　　　　電磁氣的電波（→電磁波）
(cosmic microwave background radiation) played an important
　（宇宙微波背景輻射）　　　　　　　　　扮演重要的角色
part in their studies. ⑦CMB is a very old electromagnetic wave
like a fossil sent out soon after the universe was born and it
有如化石一般　放射
reaches the earth from all around the space even today.

⑧NASA (National Aeronautics and Space Administration)
　　美國太空總署（美國國家航空暨太空總署）
launched a microwave observatory satellite to monitor CMB.
發射　　　微波觀測衛星　　　　　　　　　　為了觀測
⑨The results of analysis show that the universe was born about
　　分析的結果
13.7 billion years ago. (see p.22)
137億年前

問1　宇宙はいつ、どのようにしてできたの？

答 ①宇宙の成立については、さまざまな仮説が、立てられてきました。②その中で
　　　宇宙　形成　　　　　　　　　　　　　　　　　　　假説

現在一番広く受け入れられているのが「ビッグバン理論」です。
　　　　　　　　　　　　　　　　　　　　　　　　　　理論

③このビッグバン理論では、宇宙の始まりはとてつもなく**高温、高密度**で限りなく小
　　　　　　　　　　　　宇宙　　　　　　　　　　　高溫　高密度

さい、点のようなかたまりです。④**これがあたかも爆発したかのように急激に時間と**
　　　　　　　　　　　　　　　　　　　　　　爆炸　　　　　　　　　極速　時間

空間の拡張を始め、宇宙が生まれたとされています。
空間　擴張

⑤**宇宙誕生の時期**についても**さまざまな研究**がなされています。⑥この研究に貢献した
　宇宙誕生　時期　　　　　　　　　　　　　　研究　　　　　　　　　　　　　貢献

のは**CMB（宇宙マイクロ波背景放射）**という**電磁波**でした。⑦CMBは宇宙が誕生して
　　　　　　　　　　　　　　　　　　　　　　電磁波

間もない頃に**出された化石**のような電磁波であり、現在でも宇宙のあらゆる方向から
　　　　　　　　化石　　　　　電磁波

地球に届いています。
地球

⑧**NASA（米航空宇宙局）**はCMBを**観測**するために、**マイクロ波観測衛星**を打ち上げま
　　　美國太空總署　　　　　　　観測　　　　　　　　　　　　観測衛星

した。⑨その結果の解析から、宇宙は約**137億年前**に生まれたとされています。
　　　　　　結果　分析

問1　宇宙是在什麼時候、經由什麼方式形成的呢？

答 ①關於宇宙是如何形成的，已經存有各式各樣的**假說**。②這些理論當中，現在最廣為被接
受的理論就是「**大霹靂理論**」。③根據這個大霹靂理論，宇宙剛開始是個異常**高溫、高密**
度，但是非常**微小、有如點狀的團塊**。④這時，宛如發生了**大爆炸一般，時間與空間極速**地進行
擴張，宇宙就此而誕生了。⑤關於**宇宙誕生的時期**也有為數不少的**研究**。⑥而對於這個研究具有
貢獻的，是被稱為**CMB（宇宙微波背景輻射）**的**電磁波**。⑦CMB是宇宙誕生的瞬間所產生，有
如化石一般的電磁波，即使現在也從宇宙的所有方位朝著地球放射。
⑧**NASA（美國太空總署）**為了**觀測**CMB，發射了**微波觀測衛星**。⑨由觀測結果分析出，宇宙約
在**137億年前**所誕生。

Q 2 What is there at the end of the universe?

① First of all, how far into space can we see ? ② It is said
　　　　　　　　多遠　　　　　　　　　　　　　　　　　據說…
that the universe was born around 13.7 billion years ago.

③ Since the distance that light travels in a year is called a light
　　　　距離　　　　　光　前進　　　　　　　　　　　　　光年
year, the light the earth catches now was given off 13.7 billion
　　　　　　　　　　　　　　　　　　　　被釋放
years ago by stars 13.7 billion light years away. ④ Therefore we
　　　　　　　　　距離137億光年　　　　　　　　　因此
should be able to see 13.7 billion light years away.
應該可以

⑤ In 1929, an American astronomer, Edwin Hubble, found that the
　　　　　美國天文學家　　　　艾德溫·包威爾·哈伯　發現
universe is continuing to expand. ⑥ He said that all the stars in
　　　　　持續地擴張
the universe are moving away from the earth, and that the further
　　　　　　從…持續地遠離　　　　　　　　　　　越遠
a star is from the earth, the faster it moves.

⑦ To sum up, if we think of the end of the universe as the furthest
因此　　　　　假設　　宇宙的盡頭　　　　　最遙遠的一點
point we can see, it is 13.7 billion light years away now and it is
still extending. ⑧ When we think of the end of the universe as the
limit of space, not as the furthest point we can see, we can not
空間的界線
even find yet whether there is an end to it or not.
甚至連…都…

問2　宇宙の果てはどうなっているの？

答　①まず、宇宙はどこまで見えるかを考えてみましょう。②宇宙が誕生したのは約

137億年前**といわれています。**③**光が1年間に進む距離を1光年**といいますので、**137**

億光年離れた星が137億年前に発した光が、今やっと地球に届いていることになりま

す。④現在の地球からどこまで見渡せるかを考えると、137億光年の彼方まで、という

ことになります。

⑤1929年には、**アメリカのエドウィン・ハッブル**という学者が、宇宙は膨張し続けて

いることを**発見しました。**⑥銀河のすべての星が地球から**遠ざかるように動き**、その

速度は地球からその星までの距離に比例して**速くなる**というのです。

⑦**つまり、宇宙の果てが**「見える限界」**と考えると、現在は137億光年彼方であり、**

刻々と広がり続けていることになります。⑧見える限界ではなく、**空間としての限界**

ということになると、今のところ宇宙の果てが存在**しているかどうかさえ、**わかってい

ません。

問2　宇宙的盡頭究竟是什麼？

答　①首先，我們來思考宇宙可見的最遠距離吧！②**我們已經知道**宇宙約誕生在137億年前。
③因為光在1年之間所**行走的距離**，稱之為**光年**，也就是說距離**137億光年**的星球，在137億
年前所發出來的光，現在才終於抵達了地球。④所以說，由現在的地球能看到多遠的距離，也就
是指能看到137億光年的彼方。⑤西元1929年，美國的天文學家艾德溫・包威爾・哈伯(Edwin
Powell Hubble)發現，宇宙還在持續地膨脹與擴張。⑥銀河的所有星體，以**遠離地球的方式運
動**，移動的速度相對於地球與星體之間的距離，越遠速度越快。⑦也就是說，如果我們將**宇宙的
盡頭**定義為我們可以見到的最遙遠的一點，這是指遙遠的137億光年的彼方，而且它還不斷地持
續擴張。⑧當我們將宇宙的盡頭當作一個**空間的界線**，而不是我們可以見到的最遙遠的一點，我
們**甚至**連它是不是有盡頭都無法得知。

Q3 Is there any planet human beings can live on in space?

A ① A planet on which human beings can live must meet
人類可以居住的星球 符合
certain conditions, for example it must have water, it
條件
must have a temperature similar to that of the earth, it must be
氣溫與地球相近
made up of rock and metal, and so on. ② In the solar system Mars
由⋯所形成 在太陽系中 火星
meets some of these conditions. ③ However, the average
平均氣溫
temperature on Mars is low, around minus 60 degrees Celsius,
攝氏負60度
and about more than 90 percent of its atmosphere consists of
大氣 由⋯所組成
carbon dioxide. ④ Therefore human beings can not live on Mars.
二氧化碳
⑤ Then, how about outside this solar system? ⑥ The earth is a planet
在⋯範圍之外
moving around the sun, a fixed star. ⑦ Apart from the sun, there are
恆星 ⋯以外
countless fixed stars which give off light on their own out there in
無數的 釋放⋯ 在整個宇宙當中
the universe. ⑧ It is likely that there are some planets which have a
非常有可能
similer environment to that of the earth around those fixed stars.
與地球相同的環境
⑨ For example, there is a planet named Gliese 581 d, which moves
葛利斯 581 d
around a fixed star, Gliese 581, 20.4 light-years away from the
距離20.4光年
solar system. ⑩ Scientists think that Gliese 581 d may have a
similar environment to that of the earth because of the nature of
葛利斯581的性質
Gliese 581, which is similar to our sun, and also because of its

distance from the sun.

問3　宇宙に人間が住める星はあるの？

答 ①**人間が住める星**には、水がある、**気温が地球に近い**、岩石や金属などでできているなど、いくつかの**条件**があります。②**太陽系**でこの条件に近い星は**火星**です。③しかし、**平均気温がマイナス60度**前後と低く、**大気**の90パーセント以上は二酸化炭素です。④これでは、人間が住めそうにもありません。

⑤では、太陽系**以外**ではどうでしょうか？　⑥地球は、**恒星**である太陽の周りを回っている惑星です。⑦広い宇宙には、太陽のほかにも、自ら光を放つ星である恒星が**無数**に存在します。⑧その周りには、**地球と同じような環境**の惑星が存在する可能性は十分にあります。

⑨たとえば、太陽系から**20.4光年**離れたグリーゼ581という名前の恒星の周りを回る惑星、**グリーゼ581d**です。⑩太陽にあたる**グリーゼ581の性質**や、そこからの距離で、グリーゼ581dは地球に近い環境を持つ可能性があるとされています。

問3　宇宙中有人類可以居住的星球嗎？

答 ①人類可以居住的星球必須有水、氣溫與地球相近、由岩石或金屬所形成等，具有若干個**條件**。②在太陽系中與此條件相似的星球為**火星**。③但是平均氣溫為攝氏負60度前後、大氣有90%以上為二氧化碳組成。④因此，人類不可能居住在火星上。⑤但是，太陽系以外的行星如何呢？⑥地球是沿著身為**恆星**的太陽周圍環繞的行星。⑦在寬廣的宇宙當中，除了太陽以外還**存在著無數**個能夠**自體發光**的恆星。⑧在它們的周圍，十分有可能存在著**與地球擁有相同環境**的行星。⑨舉例來說，環繞在**距離**太陽系20.4光年、被命名為葛利斯 581的一顆恆星周圍的行星-葛利斯 581 d。⑩葛利斯581的**性質**與太陽相似，而與地球相似距離的581 d，被認為可能擁有近似於地球的環境。

Q 4 Why is the sun burning all the time?

① Like the other fixed stars, the sun also gives off large
<small>恆星</small> <small>釋放</small> <small>大量的</small>
amounts of energy and light by a nuclear fusion reaction
<small>由…產生核融合反應</small>
of hydrogen to form helium. ② The sun is about 109 times the
<small>氫</small> <small>氦</small> <small>109倍</small> <small>直徑</small>
diameter of the earth and it is thought that about 90 percent of
<small>雖然…被推測</small>
the mass of the sun is hydrogen. ③ Scientists say that as much as
<small>質量</small> <small>約…的量</small>
about 700 million tons of hydrogen is used for the nuclear fusion
<small>七億噸</small>
reaction per second at the center of the sun.
<small>每秒</small>

④ However, the sun will not go on shining forever like today.
<small>持續發光</small>

⑤ When there is little hydrogen left and helium comes to the center
<small>氫含量逐漸減少</small>
of the sun, the nuclear fusion reaction of hydrogen outside speeds
<small>加速</small>
up and expands outward. ⑥ It is called a red giant. ⑦ After that,
<small>向外側擴張</small> <small>紅巨星</small>
those gases flow out and it becomes a white dwarf, which has no
<small>流出</small> <small>白矮星</small>
nuclear fusion reactions, and it then takes billions of years to
<small>數十億年</small>
become cold.

⑧ The sun, which is thought will exist for around 10 billion years,
<small>存在</small> <small>100億年</small>
is now about 4.7 billion years old. ⑨ Therefore the sun will shine as
<small>47億歲</small>
it does today for billions of years more. (see p.23)

問4　太陽はどうして、ずっと燃えているの？

答 ①太陽もほかの**恒星**と同じように、**水素**が**ヘリウム**に変化する**核融合反応**によっ
　　氫　　　　轉變　　　核融合反應
て大量のエネルギーを発生し、光を放っています。②**直径**が地球の約**109倍**もある巨大
大量　　　　　　　　　　　　　　　　　直径　　　　　　　　　巨大
な太陽ですが、その質量の90パーセント程度が水素だと考えられています。③中心で
　　　　　　　質量　　　　　　程度　氫元素　　　　　　　　　　　　中心
は、毎秒約**7億トン**もの水素が核融合反応によって消費されているといわれていま
　　　　　　　　　　　　　　　　　　　　　　消耗
す。

④しかし、太陽はずっと今のように**輝き続ける**わけではありません。⑤**水素が残り少な**

くなると、ヘリウムが中心に集まり、その外側にある水素の核融合が**加速**して外側に
　　　　　　　　　　　　　　　　　外側　　　　　　　　　　　　加快
広がっていきます。⑥これが**赤色巨星**です。⑦その後、ガスが**流出**して核融合反応が起
　　　　　　　　　　　紅巨星　　　　　　　　　　　　流出
こらない**白色矮星**となり、**数十億年**かけて冷えていきます。
　　　　わいせい
　　　　白矮星

⑧寿命**100億年**前後と考えられている太陽は、現在約**47億歳**です。⑨そのため、**あと**
壽命　　左右

数十億年は、今の状態で輝き続けることでしょう。
　　　　　　　狀態

問4　太陽為什麼永遠持續地燃燒？

答 ①太陽與其他的**恆星**相同，藉由把氫轉變為氦的**核融合反應**，產生大量的能量、持續地發
光。②雖然巨大的太陽**直徑**約為地球的109倍，其質量的90%是為氫元素。③據說，太陽的
中心點每秒約消耗**七億噸**的氫進行核融合反應。④但是，太陽卻無法永遠持續著如同今日的**光
亮**。⑤因為氫含量逐漸減少、氦往中心聚集，位於外側的氫的核融合**反應**速度加快、並往**外側**擴
張。⑥這稱之為**紅巨星**。⑦之後，氣體**流出**、形成無法進行核融合反應的**白矮星**，經過數十億年
時間冷卻。
⑧推測壽命約在**100億年**左右的太陽，現在約為**47億歲**。⑨因此，接下來的**數十億年**間，太陽還能
持續著現在的狀態發光吧。

Q 5 Why is the earth moving around the sun?

①Everything in this universe has the power to pull all
牽引的力量
other things, and the things which has a larger mass
任何其他物體   質量越大…也就越大
have stronger power. ②This is called the universal gravitation.
萬有引力

③Put a strong magnet on the ground and roll an iron ball near the
強力的磁鐵   讓…滾動 鐵球
magnet. ④The iron ball will not go straight because the strong

magnet will pull it. ⑤The iron ball will change direction a little or
稍微改變方向
go straight to the magnet depending on the strength of the magnet
朝著…前進   藉由…   …的強弱
and the distance between these two things.
…之間的距離

⑥The iron ball may move around the magnet under certain
可能 繞著…的周圍轉動   在某整特定的條件下
conditions, such as the weight of the iron ball, its speed and the
例如 …的重量   它的速度
strength of the magnet. ⑦Planets such as the earth and others in
行星
the solar system can be thought of in this way.
太陽系   就如同這樣

⑧The earth moves around the sun because of the gravitational
因為…   …的引力
pull of the sun and the inertia which
慣性
the earth itself has: a property that
持續自我轉動的特性
keeps itself moving.

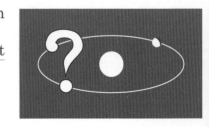

問5　地球は太陽の周りを、どうして回っているの？

答　①この宇宙では、どんなものにも、**ほかのものを引きつける力**があり、その力は

質量が大きくなるほど、大きなものになります。②これが**万有引力**です。

③**強力**な**磁石**を置き、その近くで**鉄の玉**を**転がす**とします。④鉄の玉は磁石の方向に引

っ張られ、まっすぐに転がらないでしょう。⑤磁石の**強さ**や鉄の玉と磁石の**距離**によ

って、**方向を変えて転がって**いってしまったり、そのまま磁石に引き寄せられたりし

ます。

⑥鉄の玉の**重量**と、磁石の強さ、転がした**速度**によって、**磁石の周りを回る**動きを見

せる場合もあります。⑦これが地球などの**惑星**と考えていいでしょう。

⑧地球は、地球自体が持つ**慣性**、つまり**動き続けようとする性質**と、太陽の**引力**によ

って公転しているのです。

問5　地球為何繞著太陽旋轉？

答　①在宇宙之中，不論**任何物體都有牽動其他物體的力量**。**質量越大**，這股力量也就越大。②這稱之為**萬有引力**。

③放置一塊**磁鐵**，在附近**滾動**一顆**鐵球**。④鐵球受到磁力的方向所吸引，是不是無法筆直地滾動呢？⑤依照磁力的**強弱**或是磁鐵與鐵球之間的**距離**，鐵球可能**改變方向**，或是朝著磁鐵的方向前進。

⑥根據鐵球的**重量**、磁鐵的強弱或滾動的**速度**等等，可能可以看到鐵球繞著磁鐵的**周圍轉動**。

⑦這就可以比擬為地球等等的**行星**。

⑧地球擁有自己的**慣性**，也就是**持續自我轉動的特性**，以及因為太陽的**引力**而持續地公轉。

The universe was born 13.7 billion years ago by an explosion called the Big Bang.

宇宙は137億年前、爆発（ビッグバン）によって生まれた
宇宙在137億年前因為大爆炸而誕生

The early stages of the universe
宇宙の初期のころ
宇宙形成的初期

The universe in the early stage is thought to have been a red-hot and high-density dot.

宇宙の初期は高温、高密度な一点だったと考えられている。

宇宙的初期被設想成為一個高溫、高密度的一點。

The Big Bang
ビッグバン
大霹靂

The early universe as a dot began to extend as if it had blown up.

小さな点だった初期宇宙が、あたかも爆発のように大きく膨張し始めた。

身為極微小一點的初期宇宙，宛如爆炸一般開始膨脹與擴張。

The end of the universe is 13.7 billion light years away.

宇宙の果ては137億光年の彼方
宇宙的盡頭是137億光年的彼方

Electromagnetic waves such as light given off 13.7 billion years ago, at birth of the universe, 13.7 billion light years away reach the earth now.

137億年前（宇宙の誕生時）に、137億光年の彼方で発生した光などの電磁波が、現在の地球に届いている。

137億年前（宇宙誕生初期），在137億光年的彼方產生的光等等的電磁波，現在抵達了地球。

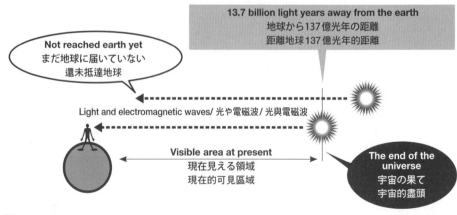

13.7 billion light years away from the earth
地球から137億光年の距離
距離地球137億光年的距離

Not reached earth yet
まだ地球に届いていない
還未抵達地球

Light and electromagnetic waves/ 光や電磁波 / 光與電磁波

Visible area at present
現在見える領域
現在的可見區域

The end of the universe
宇宙の果て
宇宙的盡頭

The sun will burn out some day

太陽はやがて燃え尽きる
太陽最終會燃燒殆盡

Today's sun
現在の太陽
現在的太陽

Red giant
赤色巨星
紅巨星

When there is little hydrogen left, the nuclear fusion reaction of hydrogen outside speeds up and expands outward.

水素が残り少なくなると、外側にある水素の核融合が加速して外側に広がる

當氫元素逐漸地減少，位於外側的氫元素核融合反應加速且向外擴張。

Giving off gases
ガスを放出する
釋放氣體

The center of it contracts by its own gravity and the gases covering it flow out.

中心は自らの重力で収縮し、外側を覆うガスは離れていく

中心因為本身的重力收縮，外側覆蓋的氣體逐漸釋放

White dwarf
白色矮星
白矮星

The gases which covered it flow out and the center of it becomes as small as the earth and a white dwarf.

外側のガスがなくなり、地球ほどの大きさになった中心部が白色矮星となる

外側的氣體逐漸流失，變成有如地球般大小的中心部位，形成白矮星

Q 6　Why is the earth rotating?

① Gases, dust and small planets were moving around
like "an eddying current" when the sun was born.
② "The eddying current" got bigger with those dust and planets
continuing run into one after another and the earth and the other
planets of the solar system, were born.
③ When two balls run into each other, they sometimes rotate.
④ This happens when the two balls do not collide, head-on, but just
brush each other. ⑤ The earth came to rotate one time in about 23
hours 56 minutes 4.06 seconds as it does today after continuing to
run into other planets and so on again and again.
⑥ A rotating ball will stop after a while because of the friction
between the ball and the ground. ⑦ However, the earth floating in
the universe, which is almost a vacuum, has no friction to stop it
from rotating. ⑧ Therefore the earth keeps rotating as long as
there is no collision with great force, such as a planet. ⑨ There are
in fact some forces which impede the rotation of the earth such as
the friction between the tides and the sea bottom among other
things. ⑩ As a result, the rotation of the earth slows down about
one over one hundred thousand seconds every year.

問6　地球はなぜ、自転しているの？

答 ①太陽が誕生したとき、ガスや塵、小さな惑星が渦を巻いて回っていました。
誕生　　　　　　　塵埃　　　　　　　　漩渦

②それらが衝突を繰り返しながら大きくなり、地球をはじめとする太陽系の惑星が生
撞擊　　反覆地進行

まれました。

③二つのボールがぶつかると、回転することがあります。④真正面からぶつからず、
旋轉　　　　　　　　　　　正面

中心からはずれてかすったような場合です。⑤地球も、惑星などとの衝突を繰り返す
中心　　　　　　　　　　　　　　　　　　　　　　　　衝撞

うちに、現在のように約23時間56分4.06秒周期で自転するようになりました。
現在　　　　　　　　　　　周期　自轉

⑥ボールは回転しても、地面に接している面で抵抗があるため、やがて止まってしま
地面　　　　　抵抗

います。⑦ところが、真空にほぼ近い宇宙に浮かぶ地球に対しては、回転を妨げる抵
真空　　　　　　　　　　　　　　　　　　　　　　　阻止

抗がありません。⑧そのため、惑星との衝突といった大きな力が加わらない限り、回
撞擊

転し続けます。⑨ただし、地球にも、潮の干満と海底との摩擦など、自転を妨げる力
漲退潮　海底　　摩擦

が働いています。⑩そのため、1年間に約10万分の1秒ずつ自転速度が遅くなっていま
自轉的速度

す。

問6　地球為什麼會自轉？

答 ①太陽誕生的時候，氣體、塵埃以及小行星，以漩渦的方式旋轉。②這個撞擊一面反覆地進行、一面擴大，便形成了地球以及其他行星所組成的太陽系。③當兩個球體撞擊時，通常會產生旋轉。④這發生在並非正面的撞擊，而是僅僅擦過邊緣的情況。⑤地球在與各個行星的反覆衝撞之中，形成了現在以23小時56分4.06秒的周期不斷地自轉。⑥球體即使旋轉，因為與接地面互相摩擦抵抗，最終會停止不動。⑦然而，對於漂浮在接近真空的宇宙中的地球而言，沒有摩擦抵抗阻止它旋轉。⑧因此，除非施加如同行星撞擊般強大的力道，地球將會不停地自轉。⑨但是，地球本身也會因為如漲退潮與海底間的摩擦等，有種種妨礙自轉的力量運行。⑩其結果，導致自轉的速度每年以10萬分之1秒的速度變慢。

Q7 Why are the earth, the sun, and the moon spheres?

① Almost all the stars, including the earth, the sun, and
　　幾乎所有的…　　　包含…
the moon, are spherical in shape. ② This is the result of
　　　　　　球的形狀　　　　　　　　因…所致
gravity.
重力
③ Various substances floating in the universe are put together by
　各式各樣的　物質　漂浮在宇宙間　　　　　聚集
gravity and a star is born. ④ In any case, gravity pulls things
　　　　　　　　　　　　　在任何情況下
toward its center. ⑤ If a star had the shape of a cube, its vertexes
　　　　　　　　假設…的話　　　　　　正方體　　頂點
would have weaker gravity as they are far from its center and the
　　　　　較弱的　　　　當…
part around its center would have stronger gravity as it is near its
　　　　　　　　　　　　　　　較強的
center. ⑥ In a sphere, the distance between its center and any
　　　　　球體　　在…之間的距離
points on its surface is the same. ⑦ Therefore any points on its
任何一點　　表面
surface have the same gravity. ⑧ Some newborn stars are not
　　　　　　　　　　　　　　　剛誕生的
spherical in shape, but even such stars gradually become spheres
　　　　　　　　　即使…　　　　逐漸地
because of gravity in the course of time. ⑨ Some minor planets
　　　　　　　隨著時間的經過　　　　　　小行星
with low mass do not become perfect balls because of their weak
　　質量小
gravity. ⑩ The minor planet named Toutatis, which came
　　　　　　　　　　　　　　　Toutatis
dangerously near to the earth in November 2008, is only 4.9 km
危險地接近…
long even at its longest point. ⑪ It was oval in shape, resembling
　　　　　最長部分　　　　　　　　橢圓形的　　　類似…
the shape of two combined stars, by observed through a telescope.
　　　　二個星體組合而成　　　　　　　　　　　望眼鏡
(see p.34)

問7　なぜ、地球も太陽も月も丸いの？

答 ①地球、太陽、月に限らず、**ほとんどの星は球の形**をしています。②これは、

重力によるものです。
重力

③重力によって宇宙をさまようさまざまな物質が集まり、星が生まれます。④重力は、
　　　　　　　　　宇宙　　　　　　　　物質

常に中心にむかって働きます。⑤仮に星が**立方体**だとすると、中心から遠い**角**の部分
　　　中心　　　　　　　　　　　正方體　　　　　　　　　　　　　　　　　角落

は重力が**弱く**なり、中心に近い部分の重力が**大きく**なります。⑥**球**では、中心から

表面のあらゆる地点までの**距離**が同じです。⑦そのため、表面のどの地点にも同じよ
表面　　　　　　地點　　　距離

うに重力が働きます。⑧**生まれたばかりの星は球体でない**こともありますが、**時間が**

たつにつれ、重力によって球体になっていくのです。

⑨**質量が小さい小惑星**などの中には、重力が小さいために球体になりきれないものも
　　質量　　　　小行星　　　　　　　　　　　　　　　　　　　　球體

あります。⑩2008年11月に地球に**異常接近**した小惑星**トータティス**は、一番長い部分
　　　　　　　　　　　　　　異常接近

でも４.９キロメートルしかありません。⑪**望遠鏡**がとらえたその姿は、**楕円形**に近
　　　　　　　　　　　　　　　　望眼鏡　　　　　　　　　　　橢圓形

い、二つの星が合体したようなものでした。
　　　　　　組合

問7　為什麼地球、太陽、月亮都是圓的？

答 ①不只地球、太陽及月亮，**所有的星體皆會形成球的形狀**。②這是因為**重力**的原因所致。③**重力**將**漂浮**在宇宙間的**各種物質**聚集，因而誕生了星體。④重力總是將物質往中心牽引。⑤假設星體是一個**正方體**，遠離中心的四個角落重力將逐漸**減弱**，而越靠近中心的部位重力則越大。⑥若是為**球體**，則由中心到**表面**的**任何一個地點**，距離皆相同。⑦因此，在表面的任何一個地點都存在著相同的重力。⑧雖然有些**剛誕生**的星球並非為球體，但隨著時間的經過、受到重力的影響將逐漸地形成球體。

⑨在質量小的小行星等等之中，由於重力小的因素，無法形成均勻的球體。⑩在2008年11月，異常接近地球的4179號小行星**Toutatis**，**最長部分只有4.7公里長**。⑪由望眼鏡所捕捉到的影像，接近橢圓形、且看似由二個星體所組合而成。

Q 8　Why don't people and things on the other side of the earth fall off even though they are upside-down?

①The word "fall" expresses a movement from top to bottom. ②The South Pole of the earth is at the bottom of a model world globe, under which there is a pedestal. ③Even if you try to put a small doll on the South Pole of a model world globe, it will fall to the pedestal.

④However, this is the case of a model world globe, not the case of the real globe. ⑤This is why there is no concept of top and bottom in space.

⑥Now, let us think about what the word "bottom" means. ⑦You will soon find that we use "bottom" to express "the direction toward the ground". ⑧To be exact, the direction of gravity to pull things is the "bottom". ⑨Anywhere on the earth, gravity pulls things toward the center of the earth.

⑩"Falling" means a movement of things toward the center of the earth by the pull of gravity. ⑪In other words, everything falls toward the center of the earth even at the South Pole though it is at the bottom of the world globe.

問8　なぜ、地球の反対側の人や物は上下逆さなのに落ちないの？

答　①私たちにとって、「落ちる」というのは、上から下に落ちることを意味します。②地球儀を見ると、南極は一番下のほうにあり、その下には地球儀を置いてある台座があります。

③地球儀の南極に小さな人形を置こうとしても、下の台座のほうに落ちてしまうでしょう。

④しかし、これは机に置かれた地球儀であって、**本物の地球**となると話が違ってきます。

⑤宇宙には、上も下もないからです。

⑥ここで、「下」とは何なのか考えてみましょう。⑦「下」というのは「**地面がある方向**」であることに気づくでしょう。⑧ちょっと難しい言い方をすれば、**重力が働く方向**が「下」です。

⑨地球上のどこでも、重力は地球の**中心付近**にむかって働いています。

⑩「落ちる」とは、重力に引っぱられ地球の中心の方向に**むかって移動する**ことです。⑪つまり、地球儀の下にある南極であっても、ものは地球の中心にむかって落ちるのです。

問8　為什麼地球另一側的人或事物，上下顛倒卻不會掉落呢？

答　①對我們來說，所謂的「掉落」指的是**由上往下墜落**的意思。②我們看地球儀時，**南極**位在**最底端的位置**，下面是放置**地球儀的台座**。③若是在地球儀的南極放置小型的人偶，就會朝底下台座的位置掉落吧！

④但是，這是指放在桌子上的地球儀而言，以**真正的地球**來說就完全不一樣了。⑤這就是為什麼，在宇宙中沒有所謂的上下之分。⑥在這裡，讓我們來思考「下」代表什麼意義吧！⑦你會意識到所謂的「下」，指的是「**地面所在的方向**」對吧！⑧用一個比較困難的說法，**重力作用的方向就是所謂的「下」**。⑨在地球上的**任何一個地方**，重力皆朝著**地球中心的附近**作用。⑩所謂的「掉落」，指的是受到重力牽引、朝著地球中心的方向**作用**這件事。⑪**也就是說**，即使物品位在地球儀下方的南極，還是會朝著地球的中心方向墜落。

Q 9 Why does the moon change its shape?

①The moon is the nearest star to the earth of all the stars
(最靠近的星體)
in solar system and it is the only satellite of the earth,
(唯一的衛星)
which goes around it. ②The moon itself does not give off light, but
(釋放…)
just seems to shine. ③The surface of the moon is covered with rocks
(只是看起來有發光)   (…的表面)   (覆蓋著…)
which reflect sunlight in different directions and some of the light
(反射)   (太陽光)   (朝著不同的方向)
reaches the earth. ④That is why the moon is not as bright as the sun.
(這就是為什麼)   (沒有如同…)
⑤Sunlight is parallel and strong. ⑥Therefore only half of the surface
(平行的)
of the moon reflects sunlight.

⑦It takes about four weeks for the moon to go around the earth one
(這個需耗費…)
time. ⑧While it goes around, the positions of the sun, the earth, and
(…的位置)
the moon change. ⑨According to the positions, the moon seen from
(根據…)   (從地球所見)
the earth changes its shape from new moon to half moon, and to full
(改變它的形狀)   (新月)   (半月)   (滿月)
moon.

⑩On the day when the moon and the sun are in the same direction
(從相同的角度觀看…)
seen from the earth, the part of the moon lit by the sun can not be
(…的部分)   (太陽光所照射)   (不能從…觀看)
seen from the earth. ⑪In this case, only the outline of the moon or
(在這種情形下)   (只有…的輪廓)
a very young moon can be seen. ⑫This is called a new moon.
(非常細小的月亮)
⑬The moon moves about 180 degrees in two weeks and, in this case,
(180度)   (在2周以內的時間)
only the part of the moon lit by the sun can be seen from the earth.

⑭This is called a full moon. (see p.35)

問9　なぜ、月の形は変化するの？

答 ①月は太陽系の中でもっとも地球に近く、地球の周りを回っている唯一の衛星です。②月は
　　　太陽系　　　　　　　　　　　　　　　　　　　　　　　　　唯一　衛星

自らが光っているのではなく、光っているように見えるだけです。③でこぼこした岩石で覆われた表
　　　　　　　　　　　　　　　　　　　　　　　　　　　　　岩石　表面覆蓋

面が太陽の光を乱反射し、その一部が地球に届くのです。④月が太陽ほど明るくないのは、このた
　　　　　反射　　　　　　　　　　抵達

めです。

⑤太陽光は、平行で強い光です。⑥そのため、太陽の光を受けて光って見えるのは、月の半分だけ
　　太陽光　　平行　　　　　　　　　　　　　　　　　　　　　　　　　　　　　一半
です。

⑦月は地球の周りを約4週間かけて1回りします。⑧その間に、太陽、地球、月の位置関係は変わり
　　　　　　　　　　　周　　　　　　　　　　　　　　　　　　　　　　　相關位置
ます。⑨この位置関係の変化により、地球から見た月は新月、半月、満月と形を変えるのです。
　　　　　　　　　　　　　　　　　　　　　　　新月　半月　滿月　　變化

⑩地球から見て月と太陽が同じ方向にある日は、太陽光が当たっている部分が見えません。⑪地球
　　　　　　　　太陽　　方向　　　　　　　　　　　　　　　　　　　　　　　　　　　　　　地球

からは月の輪郭、あるいはごく細い月がわずかに見えるだけです。⑫これが「新月」です。
　　　　輪廓　　　　　　　　　　　　　　　　　　　　　　　　　　　　　　新月

⑬2週間ほどたつと月は約180度移動し、地球からは月の太陽が当たっている部分しか見えなくな
　　　　　　　　　　　　　　移動

ります。⑭これが満月です。
　　　　　　滿月

問9　為什麼月球的形狀會改變呢？

答 ①月球是太陽系當中最靠近地球，唯一在地球周圍環繞的一顆衛星。②月球本身並不會發光，
是看起來有發光的樣子。③表面覆蓋的凹凸不平的岩石，將太陽光任意反射，其中一部分抵達
了地球。④月球沒有像太陽一樣明亮，就是這個原因。⑤太陽光是平行且直射的強光。⑥因此，月球只
有一半能夠接受且反射太陽光。⑦月球繞地球一周需耗時約4周。⑧在繞行的期間，太陽、地球、月球
的相關位置將會變化。⑨因為這個相關位置的變化，從地球所見的月球就會有新月、半月、滿月的形
狀變化。⑩若是從地球觀看到月亮與太陽位在同一個方向，我們看不見太陽光所照射的部分。⑪從地
球只能看見月球的輪廓，或是非常細小的月亮。⑫這稱之為「新月」。⑬經過約2周的時間，月球約180
度移動，由地球只能觀看到月亮正對太陽的部分。⑭這稱之為滿月。

Q10 Why does the moon keep up with us when we are walking?

① Suppose that you are walking along the street at night,
假設…

and that, by the roadside, there is a telephone pole 10
在路邊　　　　　　　　　電線杆

meters ahead and an iron tower one kilometer ahead. ② When you
10公尺的前方　　　鐵塔

walk 10 meters, you are just beside the telephone pole. ③ At that
正好在…的側邊

time, the iron tower is 990 meters ahead of you and it looks almost
看起來幾乎像…

like what you saw 10 meters behind. ④ When you are moving, you
你看到的…　　10公尺的後方

feel that distant things do not change their position as much as
遠處的事物　　並非…　改變位置

nearby things.
近處的事物

⑤ It is as far as 380 thousand kilometers from the earth to the
約…的距離　380,000公里

moon. ⑥ Even if you move a few kilometers, you do not feel that the
即使…　　　　　　　　　完全

position of the moon you see changes at all.
你看到的月亮的位置

⑦ In daily life, you do not feel that the position of a distant thing

changes (a) when both you and the thing are moving at the same
相同的速度

speed or (b) when both you and the thing are standing still.
靜止的

⑧ Since you supposed that you were walking along the street at
因為…

night, it is clear that (b) is not the case here. ⑨ Therefore you get it
很明顯的…　　　不是指這個狀態　　　　產生錯誤

wrong and think that the moon keeps up with you.
跟著…

問10　歩いていると、なぜ月はついてくるの？

答　① たとえば、夜道を歩いていて10メートル前方の道路わきに電柱があり、1キロ

前方に鉄塔があったとします。② 10メートル進むと、あなたは電柱の真横にたどり着

きます。③ 一方鉄塔は、990メートル前方にあり、**10メートル前から見たのとほとん**

ど変わっていないでしょう。④ 動いているあなたから見ると、**遠くのものは近くのも**

のに比べて**位置変化**が少なく感じられるのです。

⑤ 地球から月までの距離は**380,000キロメートル**もあります。⑥ 2キロや3キロ移動し

たところで、あなたから**見た月の位置はまったく変わりません**。

⑦ 日常生活で、あなたと離れたものとの位置関係が変わらないのは、（a）両方とも**同**

じ速度で移動しているか、（b）両方とも**静止**しているかのどちらかです。⑧ あなたは

夜道を歩いている**ので**、（b）ではない**ことは明らか**です。⑨ そのため、月が**同じよう**

に**動いていると錯覚**してしまうのです。

問10　為什麼當我們走路，月亮會跟著我們移動呢？

答　① 舉例來說，夜晚獨自一個人行走，10公尺的前方路邊有根電線杆，1公里前方有座鐵塔。② 往前走10公尺，你會**正好抵達電線杆的側邊**。③ 另一方面，鐵塔位於前方990公尺處，與**先前10公尺的位置所見，是否幾乎沒有改變**呢？④ 從移動的你本身來看，**遠處的事物與近處的事物相比，比較沒有位置變化的感覺**。

⑤ 由地球到月球的距離，為**380,000公里**。⑥ 即使移動個2至3公里，從你所**觀看的月亮位置也完全不會改變**。⑦ 在日常生活當中，你與遠離中的物體相關位置不會變化有兩種情形：（a）兩方皆用**相同的速度**移動、（b）兩方皆呈**靜止**的狀態。其中一種。⑧ **因為**你在夜晚獨自一個人行走，（b）這裡很**明顯的不是**指狀態。⑨ **因此**，這是你產生了跟月亮一樣在移動的錯覺所導致。

Big stars are spherical in shape because any points on their surface have the same gravity.

大きな天体が球形なのは、表面の重力が均等に働くから
大型的天體為什麼會形成球形，是因為所有的表面重力均相等所至

Sphere
球
球形

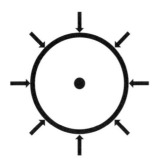

Cube
立方体
立方體

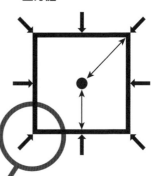

Any points on their surface have the same gravity.

表面のどの場所でも、重力は均等に働く。

所有的表面重力均相等。

The nearer the point comes to its center, the stronger gravity becomes.

中心に近いほど、大きな重力が働く。

離中心點越近，重力就越大。

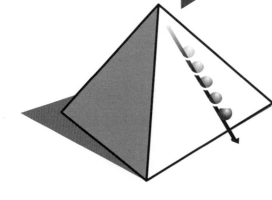

Everything is always likely to move toward the point having strong gravity from the point having weak gravity.

ものは重力が小さいところから大きいところへ移動しようとする。

任何事物皆是由重力小的地方往重力大的地方移動。

Therefore it became a sphere, which has the same gravity at any points on its surface.

その結果、全体に重力が均等な球形になる。

因此，整體就形成重力平均的球形。

The part of the moon lit by the sun changes its shape according to the positions of the moon and the earth when we see from the earth.

月で光が当たっている部分は、地球との位置関係によって見え方が変わる

太陽光照射到月球的部份，會因為地球與月球的相關位置，所見的陰影會有所改變

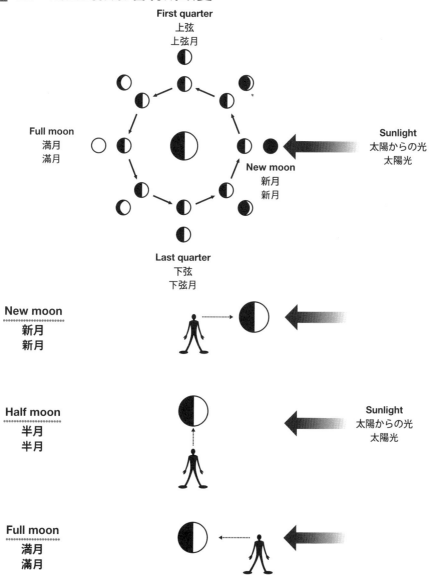

First quarter
上弦
上弦月

Full moon
満月
滿月

Sunlight
太陽からの光
太陽光

New moon
新月
新月

Last quarter
下弦
下弦月

New moon
新月
新月

Half moon
半月
半月

Sunlight
太陽からの光
太陽光

Full moon
満月
滿月

Q 11　Why do shooting stars fall?

[1] Shooting stars are called "stars", but in fact they are
流星　　　　　　　　　　　　　　　　　　　　事實上

not stars. [2] They are small dust particles floating along
灰塵微粒　　　　漂浮在…

the earth's orbital path. [3] It is said that the smallest dust particle
地球的公轉軌道

is about 0.1 millimeters across and even the biggest dust particle
直徑0.1毫米

is only several centimeters across. [4] These dust particles burn up
燃燒

and give off bright light after they enter the earth's atmosphere.
釋放…　耀眼的光芒　　　　　　　　　地球的大氣層

[5] Most of these dust particles, which could become shooting stars,
可能會形成流星

were left by comets. [6] A comet is like a mass of ice which has a lot
彗星　　　　　　　　　　　　　冰塊

of dust inside it. [7] When comets are far from the sun, they are
內部含有大量的塵埃

frozen hard. [8] However, when they come nearer to the sun, they
處於冰凍的狀態

melt little by little and leave various substances behind. [9] The
融化　一點一點地　　　殘留…　各式各樣的物質

vapor-trail-like substances left after a comet passes by are called
有如飛機雲一般的　　　　殘留　　　　　　通過

"dust trails".
彗尾

[10] Fine dust particles shower onto the earth when it goes through
細小的　　　　　　降落到…　　　　　　　　　　通過

the "dust trail" left along its orbital path.
殘留…

[11] As a result, a lot of shooting stars are
結果

seen from the earth.

問11

流れ星はどうして流れるの？

 答 ①**流れ星**は、流星と呼ばれますが星ではありません。②**地球の公転軌道**にさまよ
_{公轉軌道}

っている小さな塵です。③小さなもので**直径0.1ミリメートル**くらい、大きなものでも
_{直徑}

直径数センチメートルだといわれています。④これが**地球の大気圏**に突入するときに
_{大氣層}

燃え、**明るい光を放つ**のです。
_{はな}

⑤流星のもとになる塵は、主に**彗星**が残していったものです。⑥彗星は、**たくさんの塵**
_{すいせい}
_{彗星}

を含んだ氷の塊のようなものです。⑦太陽から離れた場所にいるときは**凍りついてい**
_{物質}_{かたまり} _{時候} _{こお} _{冰凍}

ます。⑧太陽に近づくにつれ氷が**融け**、さまざまな**物質**を残していきます。⑨彗星が**通**
_と _{物質} _{通過}

過したあとに**飛行機雲**のように残る物質は、ダストトレイルと呼ばれます。
_{飛機雲}

⑩地球が、**公転軌道**上に残されたダストトレイルの中を**通過**すると、**細かい塵**が地球
_{公轉軌道} _{通過}

上に**降り注ぎます**。⑪このとき、地球上からは多くの**流星**が**観測**できるのです。
_{流星　觀測}

問11

流星為什麼會移動？

答 ①流星，雖然稱之為流星但卻不是星星。②是在**地球公轉軌道上徘徊的灰塵微粒**。③小的有
到**直徑約0.1毫米**、大的甚至有到直徑數公尺。④它們在闖入**地球大氣層**的時候**燃燒**，釋放
出耀眼的光芒。⑤組成流星的物質主要是塵埃，多半是彗星所殘留下來的物質。⑥彗星是**含有大**
量塵埃、有如冰塊一般的物質。⑦在距離太陽遙遠的時候，是處於**冰凍的狀態**。⑧隨著越來越接
近太陽，冰塊逐漸**融化**並且**殘留各式各樣的物質**。⑨當彗星**通過**之後，會殘留下有如飛機雲一般
的物質，這稱之為**彗尾**。⑩當彗星**通過**地球，彗尾殘留在公轉軌道上，**細小的塵埃**落到了地球。
⑪這個時候，從地球上就可以**觀測**到許多的流星。

Q12　Why do stars shine?

① Of all the stars, the ones which give off light are called
在所有的星體之中　　　　　　　釋放…

fixed stars. ② Fixed stars give off energy because of
恆星

nuclear fusion reactions. ③ Nuclear fusion reactions form four
核融合反應　　　　　　　　　　　　由…形成

hydrogen atoms into one helium atom. ④ This type of reaction is
氫原子　　　　　　　　氦原子

also applied to hydrogen bombs and it gives off a great deal of
也被應用到…　　　氫彈　　　　　　　　　　　　非常巨大的

energy.

⑤ The sun is one of the fixed stars. ⑥ It is said that the temperature
　　　　　　　　　　　　　　　　　　　　　　據說…　　　　溫度

of the corona around the sun is more than one million degrees C. ⑦
太陽周圍的日冕　　　　　　　　　　　　　　攝氏100萬度

Fixed stars shine because they are heated to a high temperature
　　　　　　發光　　　　　　　　被加熱

and give off a great deal of energy such as light and heat energy.
　　　　　　　　　　　　　　　　像…這樣的　光與熱的能量

⑧ Some stars do not give off light. ⑨ Examples of these are the

planets going around a fixed star such as Jupiter, Venus, and the
　　　　在…的周圍圍繞　　　　　　　　　　木星　　金星

earth. ⑩ Seen from the earth, these two stars appear to shine.
　　　　從…來觀看時

⑪ However, these planets do not give off light but just reflect light
　　　　　　　　　　　　並非…　　　　　　　　　　　反射…

from the sun, a fixed star.

問12　なぜ、星は光るの？

答 ①星の中でも自ら光を出している星は、**恒星**と呼ばれています。②恒星は、核融
恆星

合によってエネルギーを放出しています。③**核融合**というのは、4個の**水素原子**が1個
核融合　　　　　　　　　　　氫原子

のヘリウム原子になる**反応**です。④これは**水素爆弾**に使われている反応で、非常に大
反應　　　　　　氫彈

きなエネルギーを**発生**します。
產生

⑤太陽も恒星の一つです。⑥**太陽を取り巻くコロナの温度は摂氏100万度**以上もあると
温度　攝氏

いわれています。⑦恒星が**光る**のは、このように核融合によって**高温**になり、**大量**の
高温　　　　大量

光や熱などのエネルギーを**放出**しているからです。
釋放

⑧自ら光を出さない星もあります。⑨たとえば、**木星や金星**、地球などのように、恒星
木星　金星

の周りを回る惑星です。⑩地球から見ると、これらの2つの星も光り輝いているよう

に見えます。⑪これらの惑星は実際に輝いている**わけ**ではなく、恒星である太陽の光
實際

を**反射**しているだけなのです。
反射

問12　星星為什麼會發光？

答 ①在所有的星體之中，可以自己**發出光芒**的稱之為**恆星**。②**恆星**因為**核融合**反應而能夠**釋放能量**。③所謂的核融合，是由4個**氫原子**融合成1個**氦原子**的反應。④這個反應也被應用到**氫彈**上，可以**產生**非常巨大的能量。

⑤太陽也是恆星之一。⑥**據說圍繞著太陽的日冕，溫度達到攝氏100萬度**以上。

⑦恆星所散發出來的**光**，是由這樣的核融合反應所產生的高溫、大量的**光與熱等等的能量**釋放而來的。

⑧也有無法自體發光的星體。⑨舉例來說，像**木星、金星或地球**這樣，**圍繞在恆星周圍的行星**。

⑩從地球**觀看**，這兩個行星看起來也像是在發光。⑪實際上這兩個行星並**不會**發光，是因為**反射**了身為恆星的太陽光所導致的結果。

This is how the solar system is made up.

太陽系の構成はこうなっている
這是太陽系的構造圖

The solar system
太陽系
太陽系

Black hole
ブラックホール
黑洞

Sun
太陽
太陽

Meteor
隕石
隕石

Mercury
水星
水星

Venus
金星
金星

Pluto
彗星
彗星

Earth
地球
地球

Mars
火星
火星

Jupiter
木星
木星

Space rocket
宇宙ロケット
太空船

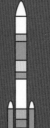

Saturn
土星
土星

Uranus
天王星
天王星

Satellite
人工衛星
人造衛星

Neptune
海王星
海王星

Universe
宇宙
宇宙

Questions about the earth

第 2 章

地球の疑問
地球的疑問

Q 13   Why does the sky look blue?

① Smoke is sometimes used for a production in a concert
煙霧　　　　　　　　　　　　演出　　　　　　　音樂會會場
hall and so on. ② When smoke fills the hall, you can see
　　　　　　　　　　　　　充滿著…
a spotlight, which you can not see in other ways, as a line of a color
聚光燈　　　　　　　　　　　　用其它的方法　　　　線
such that you can see a red spotlight as a red line. ③ This is a
就如同是…　　　　　　　　　　　　　　　　　　　　　　現
phenomenon called scattering whereby light hits particles in the
　　　　　　　　　散射　　　　　　　　碰撞到… 顆粒
air and changes direction. ④ By this scattering, light, which you
　　　　改變方向
can not see in other ways, comes to be seen as a color.

⑤ The color of a visible light, a kind of electromagnetic wave,
　　　　　　　　可視光線　　　　　　　　電磁波
changes according to its wavelength. ⑥ Light with a short
　　　　　根據…　　　　　　波長　　　　　波長短的光
wavelength looks blue and light with a longer wavelength looks
　　　　　　　　　　　　　　　　波長較長的光
yellow or red. ⑦ Scattering changes according to the wavelength of
light and the size of the particles in the air. ⑧ Smoke on the stage
　　　　　大小
scatters all the colors of light such as red, blue, and yellow because
散射…
the particles of the smoke are big. ⑨ Small particles are likely to
　　　　　　　　　　　　　　　　　　　　　　　　　　容易…
scatter blue light, which has a short wavelength.

⑩ There are molecules of oxygen and nitrogen in the thick layer of
　　　　　　分子　　　氫　　　　氮　　　　厚厚的大氣層
atmosphere covering the earth and they scatter sunlight. ⑪ They
　　　　　覆蓋著…　　　　　　　　　　　　　　太陽光
are likely to scatter blue light but do not scatter much red and
yellow light because the molecules of gas in the air are small.
　　　　　　　　　　　　　　　　氣體
⑫ The sky looks blue because blue light in sunlight is scattered in
　　　　看起來…

the air. (see p.54)

問13　空はどうして青いの？

答 ①**コンサート会場**などの**演出**で、**スモーク**が使われることがあります。②煙が会場に**充**
　　　　會場　　演出　　　　　　　　　　　　　　　　　　　煙霧　　　迷漫
満すると、それまで見えなかった**スポットライト**の光が、赤い光は赤い線になってというよう

に、色の線として見えるようになります。③これは、小さな粒にぶつかった光が**方向を変える**

「**散乱**」と呼ばれる**現象**です。④散乱によって、それまで見えなかった光が「色」として見える
　散射　　　　　　　現象
ようになるのです。

⑤**電磁波**である**可視光線**の色は、その**波長**によって変わります。⑥**波長が短い光**は、私たちに
　電磁波　　　看得見的光線　　　　　波長
青く見え、**波長が長くなる**と黄色や赤に見えます。⑦散乱は、光の波長と散乱する「粒」の**大**
　　　　　　　　　　　　　　黄
きさに関係します。⑧**ステージ**のスモークが赤、青、黄色、すべての光**を散乱する**のは、煙の粒
　　　　　　　　原故
が大きいからです。⑨小さな粒だと、波長が短い青い光ほど散乱します。

⑩地球をとり囲む**厚い大気の層**には**酸素**や**窒素**の**分子**があり、これらが**太陽の光**を散乱しま
　　　　　　　　大氣　　　　　氧　　　氮
す。⑪**気体**の分子は小さいので、青い光を多く散乱し、赤や黄色の光はほとんど散乱しませ
　　氣體　分子
ん。⑫空**が見える**のは、太陽光の中の青い光が空で散乱しているからなのです。

問13　天空為什麼是藍色的？

答 ①在**音樂會的會場**等等的**演出**，經常會有使用煙霧的情形。②當煙霧迷漫了整個會場，這
時候我們無法看見**聚光燈**的光，紅色的光會形成如同紅色的直線一般，看起來像是一條紅
線。③這是因為光**照射**到微小的**顆粒**，**改變了行進方向**，產生稱為「**散射**」的**現象**。④因為散
射，你看不到的光線變成了可見的「**色彩**」。⑤**看得見的光線**也可以說是一種**電磁波**，依**不同的**
波長而有所變化。⑥**波長短的光**，我們看起來是藍色的；**波長變長**，看起來則是黃色或紅色的。
⑦散射與光的波長，以及飄落在空氣中的顆粒有**很大的關係**。⑧舞台的煙霧能夠**散射**紅色、藍
色、黃色等等所有的光線，是因為煙霧的粒子大顆的原故。⑨小顆粒子只能散射波長短的藍色光
線。⑩圍繞在地球的**厚厚一層大氣裡**，含有**氧**與**氮**的**分子**，它們會散射太陽光。⑪因為**氣體**的分
子小，散射大部分的藍色光，紅色與黃色幾乎不散射。⑫天空**看起來**是藍色的，是因為天空散射
了太陽光之中的藍色光所導致。

Q14　Why are sunsets red?

①The atmosphere of the earth is more than 500 kilometers
大氣層　　　　　　　　　　　　　　　　　　厚度500公里

thick. ②There are big particles such as water vapor and
　　　　　　　　　　　　　顆粒　　　　　　水蒸氣

dust in the low layer of the atmosphere from the surface of the
塵埃　　底層　　　　　　　　　　　　從地表算起到35公里的…

earth to 35 kilometers high. ③It was mentioned on pp. 42-43 that
　　　　　　　　　　　　　　　同…頁所述

blue light in sunlight is likely to be scattered by small particles
　　　　　　太陽光　容易…　　因為…而散色

such as molecules of oxygen and nitrogen. ④On the other hand,
　　　　分子　　　氧　　　　氮　　　　　另一方面

bigger particles such as water vapor and dust scatter red and yellow

light.

⑤In the evening, sunlight travels a long distance in the air along
　　　　　　　　　　　　前進　　長距離

the surface of the earth and reaches your eyes. ⑥As this happens,
　　　　　　　　　　　　到達…　　　　　　當這個發生的其間

blue light is scattered and little of it can reach your eyes. ⑦
　　　　　　　　　　　　　幾乎無法…

However, sunlight in the evening looks reddish because red light
　　　　　　　　　　　　　　　　　　紅色的

travels through the air with little scattering.
　　　　　　　　　　　　　幾乎沒有散射

⑧The sky in the west becomes red

when this red light is scattered by

big particles such as water vapor

and dust. ⑨This is what red sunsets
　　　　　　　　　夕陽就是這樣產生的

are. (see p.54)

問 14　夕焼けはなぜ、赤いの？

答 ①地球の**大気圏**は**500キロメートル以上の厚さ**があります。②そのうち、**地表か**
ら高度35キロメートルまでの**低層**には**水蒸気**や**塵**など、大きな**粒**が多く漂っています。③**太陽光**の中の青い光は、大気中の**酸素**や**窒素**の**分子**など、小さな粒によって**散乱しやすい**ことは、42〜43ページで**説明**したとおりです。④**一方**、水蒸気や塵など**比較的大きな粒**は、赤や黄色の光を**散乱**します。

⑤**夕方**になると、太陽の光は**地表**に沿うように、**長い距離**、大気の中を**通過**して私たちの目に**届きます**。⑥その間に、青い光は散乱して少なくなってしまいます。⑦一方、赤い光は、ほとんど散乱することなく大気を突き抜けるので、夕方の**太陽光**は**赤っぽく見えます**。

⑧この赤っぽい光を、水蒸気や塵など、大きな粒が散乱すると、空が赤く染まったように見えます。⑨これが**夕焼けの正体**です。

問 14　夕陽為什麼是紅色的？

答 ①地球的大氣層厚度超過500公里以上。②其中，高於地表35公里的底層，充滿了水蒸氣、塵埃等等，漂浮著許多顆粒大的懸浮粒子。③如同P42〜43頁所述，太陽光之中的藍色光、因為大氣中的氧與氮的分子等等微小的顆粒而容易產生散射。④另一方面，水蒸氣與塵埃等等顆粒大的粒子，則會散射紅色或黃色的光。

⑤到了傍晚，太陽光沿著地球表面，通過長距離的大氣、射入我們的雙眼。⑥其間，藍色光幾乎散射殆盡。⑦相反的，因為紅色光幾乎沒有散射且能通過大氣層，傍晚的太陽光看起來就是紅色的。⑧當紅色光透過水蒸氣、塵埃等等大顆粒子的散射，天空看起來就像被染成了紅色。⑨這就是夕陽形成的由來。

Q15 Why is seawater salty?

①There was no sea on the earth four billion years ago.
40億年前

②At that time, the earth was extremely hot and water
極端的炎熱

was held in the atmosphere in a state of vapor. ③It is thought that
包含在… 大氣 以…的狀態 水蒸氣

the atmosphere contained hydrogen chloride besides water vapor,
氯化氫 除了…之外 水蒸氣

carbon dioxide, nitrogen, and so on.
二氧化碳 氮

④After the earth cooled down in the course of time, it started
冷卻 隨著時間經過

raining. ⑤It was the rain of hydrogen chloride, that is hydrogen
也就是說

chloride in water. ⑥This rain of hydrogen chloride fell to the
溶於水中的氯化氫 降落…

ground, dissolved various substances, and made pools in hollow
溶解… 各式各樣的物質 聚集 在低窪的地區

places. ⑦This is how seas were born.

⑧Newborn seawater was not salty like it is today, but it was sour
甫誕生的 鹹 如同今日的 酸

because it was hydrogen chloride water. ⑨The hydrogen chloride

vaporized and fell to the ground as rain again and again. ⑩During
蒸發 反覆地

that time, hydrogen chloride water dissolved a large amount of
大量的

sodium and magnesium on the sea bottom and on the ground. ⑪
鈉 鎂 海底

As a result, sodium chloride and magnesium chloride were formed
氯化鈉 氯化鎂 形成…

in a chemical reaction and accumulated in the sea. ⑫This is how
因為化學反應 蓄積在…

the sour seawater of the old days became salty as it is today.
昔日的

問 15 海水はなぜ、塩辛いの？

答 ①今から**40億年以上前**、地球に海はありませんでした。②**非常に高温**だったた
め、水分は**水蒸気の状態で大気に含まれていました**。③当時の大気は、**水蒸気、
二酸化炭素、窒素**などのほかに、**塩化水素**を含んでいたと考えられています。
④**時間がたつにつれて地球の温度が下がると、雨が降るようになりました**。⑤雨といっ
ても、**水に塩化水素が溶けた塩酸の雨です**。⑥**塩酸の雨は地表にある物質を溶かしな**
がら流れ、くぼんだ場所にたまりました。⑦これが海の誕生です。
⑧**生まれたばかりの海の水は、今のように塩辛くなく、酸っぱい塩酸でした**。⑨塩酸は
蒸発して、再び雨となり地上に降り注ぐことを繰り返しました。⑩その過程で、塩酸
が**海底や地表のナトリウムやマグネシウムなどを大量に溶かし出しました**。⑪これが
化学反応によって塩化ナトリウムや塩化マグネシウムとなり、海に蓄積されていきま
した。⑫こうして酸っぱかった海の水が、今のように塩辛くなったのです。

問 15 海水為什麼是鹹的？

答 ①距今40億年以前，地球上並沒有海洋。②當時因為氣溫非常高，水分以**水蒸氣的狀態包
含在大氣之中**。③當時的大氣除了**水蒸氣、二氧化碳、氮**等等之外，可能還含有氯化氫。
④隨著時間經過、地球的溫度下降，便形成了雨降落。⑤雖然說是雨，卻是指氯化氫溶於水中所
形成的鹽酸（氯化氫水溶液）。⑥鹽酸（氯化氫水溶液）降落後，一面溶解地表上的物質一面流
動，在低窪的地區聚集。⑦這就是海洋的誕生。
⑧甫誕生的海洋的水，並不如今日的海水一般的**鹹**，而是**酸性**的鹽酸（氯化氫水溶液）。⑨鹽酸
蒸發後，再次形成雨降落到地面，這個過程不斷地循環。⑩過程中，鹽酸不斷溶解出大量的鈉與
鎂到**海底或地表**。⑪其間因為**化學反應**，形成了氯化鈉與氯化鎂，且蓄積在海洋裡。⑫因此，過
去呈酸性的海水，變成像今日海水一般的鹹。

Q16 Why is a mountain top colder though it is nearer to the sun?

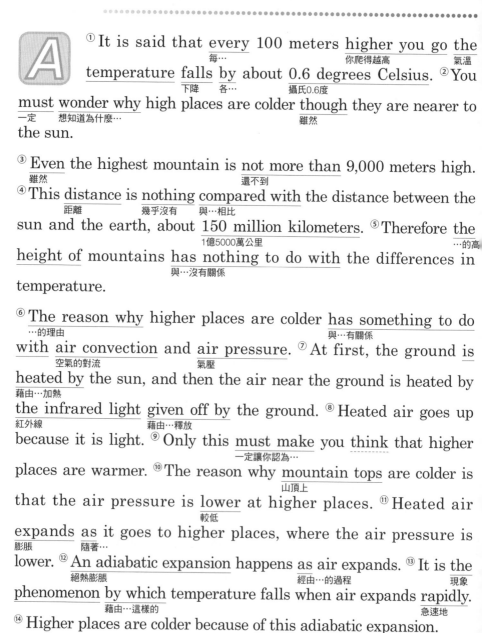

① It is said that every 100 meters higher you go the temperature falls by about 0.6 degrees Celsius. ② You must wonder why high places are colder though they are nearer to the sun.

③ Even the highest mountain is not more than 9,000 meters high. ④ This distance is nothing compared with the distance between the sun and the earth, about 150 million kilometers. ⑤ Therefore the height of mountains has nothing to do with the differences in temperature.

⑥ The reason why higher places are colder has something to do with air convection and air pressure. ⑦ At first, the ground is heated by the sun, and then the air near the ground is heated by the infrared light given off by the ground. ⑧ Heated air goes up because it is light. ⑨ Only this must make you think that higher places are warmer. ⑩ The reason why mountain tops are colder is that the air pressure is lower at higher places. ⑪ Heated air expands as it goes to higher places, where the air pressure is lower. ⑫ An adiabatic expansion happens as air expands. ⑬ It is the phenomenon by which temperature falls when air expands rapidly. ⑭ Higher places are colder because of this adiabatic expansion.

問16　山の上は太陽に近いのに、どうして寒いの？

答
① 標高が100メートル高くなるごとに、気温は摂氏約0.6度下がるといわれています。
　標高　　　　　　　　　　　　　　　　　　気温　摂氏

② 「高いほうが太陽に近いのに、どうして寒くなるの？」と、疑問に思うでしょう。
　　　　　　　太陽　　　　　　　　　　　　　　疑問

③ 高いといっても、一番高い山でも9000メートル以下。④ 太陽と地球の距離、約1億5000
　　　　　　　　　　　　　　　　　　　　最

万キロメートルに比べれば、この差はごくわずかです。⑤ これでは気温に影響しません。
　　　　　　　　　　　　　　　　　　　　　　　　　　　　　　　　　影響

⑥ 標高が高くなると気温が下がる理由は、空気の対流と気圧に関係します。⑦ まず、太陽の
　　　　　　　　　　　　　　　理由　　　　　對流　　　氣壓　關係

熱によって地面が温められ、地面から放たれる赤外線によって、地面付近の空気が温めら
　　　　　地面　あたた　　　　　　　　はな　　　紅外線　　　　　　附近

れます。⑧ 温められた空気は軽くなるため、上昇します。⑨ これだけなら、上に行くに従って
　　　　　　　　　　　　　　　　　　　　上升

気温が上がるはずです。⑩ 山の上のほうが気温が低いのは、標高が高くなるに従って気圧が
氣溫

低くなるからです。⑪ 暖かい空気のかたまりは、気圧が低い上空に行くにつれて膨らみま
　　　　　　　　　　　　　　　　　　　　　　　　　　　　　　　　　　膨脹

す。⑫ その過程で、断熱膨張という現象が起こります。⑬ 気体が急激に膨張すると、温度が
　　　　　　絶熱膨脹　　　　現象　　　　　　　　　急速　膨脹

下がる現象です。⑭ この断熱膨張によって、高度が上がるにつれ気温が下がるのです。
　　　現象　　　　　　　　　　　　　　　高度

問16　為什麼山頂接近太陽，卻比較寒冷呢？

答
① 據說標高每升高100公尺，氣溫會下降約攝氏0.6度。② 你一定會提出一個疑問：「既然高處接近太陽，為什麼比較寒冷呢？」③ 雖然高，但即使最高的山也**低於**9000公尺。④ 跟太陽與地球間相距約1億5000萬公里的距離相比，這樣的差距相當的微小。⑤ 因此這樣的高度對氣溫幾乎**沒有影響**。⑥ 標高升高、氣溫會下降的理由，是因為**空氣的對流與氣壓之間的關係**。⑦ 首先，因為太陽的熱能使地面的溫度升高，由地面**反射**的紅外線讓地面附近的空氣**變得溫暖**。⑧ 因為暖空氣較輕的原故，開始上升。⑨ 如果只有這樣向上，氣溫理所當然的會升高。⑩ **山頂上的氣溫**為什麼較低，是因為隨著標高越高，氣壓越低的緣故。⑪ 暖空氣團一面**膨脹**、一面上升到低壓的高處。⑫ 在這個過程中，產生稱為**絕熱膨脹**的現象。⑬ 也就是氣體**急速**地膨脹時，溫度會下降的**現象**。⑭ 因為這個絕熱膨脹，隨著高度越高、氣溫就會降低。

Q17　How are auroras formed?

①The sun keeps giving off plasma. ②Plasma is a stream
持續地釋放…　　　　　電漿　　　　　　　一群集合體
of particles which have a positive or a negative electric
粒子　　　　　　　　帶有正離子與負離子的電荷
charge. ③It is sometimes called the solar wind because it moves
　　　　　　　　　　　　　　太陽風
like a wind and covers the solar system.
　　　　　覆蓋著…　太陽系

④The earth is a huge magnetic body and is surrounded by a
　　　　　　　巨大的磁鐵　　　　　　　　被…包圍著
magnetic field, whose south pole is around the Arctic Zone and
磁場　　　　　　　　S極　　　　　　　　北極圈
whose north pole is around the Antarctic Zone. ⑤The earth's
　　　　N極　　　　　　　　南極圈
magnetic field is stretched widely in the opposite direction to the
　　　　　　　被大大地牽引　　　　與…相反的方向
sun by the solar wind. ⑥The sun has a magnetic field, too. ⑦The
sun's magnetic field is strong enough to cover the earth's magnetic
　　　　　　　　　　　　足以…
field.

⑧Magnetic fields have the property of keeping plasma away.
　　　　　　　　　　有…的特性　　排斥…在外
⑨However, when the sun's magnetic field touches the earth's
　　　　　　　　　　　　　　　　　　接觸到…
magnetic field, plasma enters the earth's magnetic field and rushes
　　　　　　　　　　　　　　　　　　　　　　　進入…
into the earth's Arctic Zone and Antarctic Zone along the earth's
　　　　　　　　　　　　　　　　　　　　　　地球磁場
magnetic field lines with a great deal of energy. ⑩At that time,
　　　　　　　　　　大量的…
plasma collides with atoms and molecules of oxygen and nitrogen
　　　　撞擊　　　原子　　分子　　　氧　　　氮
in the upper atmosphere and light-emitting phenomena take place.
　　高空大氣層　　　　　　釋放光　　現象　　　引起
⑪This is how auroras are formed. (see p.55)
　　極光如何形成的方式

問17　オーロラはどうやってできるの？

答 ①太陽は、**いつもプラズマを**放出**しています。②プラズマとは、**正または負の** 電荷をもった粒子の集まりです。③風のように流れて太陽系を覆っていることか ら、**太陽風**とも呼ばれています。

④地球は北極付近をS極、南極付近をN極とする**大きな磁石**のかたまりであり、**磁場に** 囲まれています。⑤地球の磁場は、太陽風の影響で太陽と反対方向に大きく引き伸ば されたような形をしています。⑥太陽にも磁場があります。⑦これは非常に強いもの で、地球の磁場を覆っています。

⑧磁場はプラズマを寄せつけない性質があります。⑨しかし、太陽の磁場と地球の磁場 が接すると、プラズマは地球の磁場に入り込み、**大きなエネルギーを持ちながら地球 の磁力線**に沿って北極や南極付近に**なだれ込みます。**⑩このとき、プラズマが**高層大 気**中の酸素や窒素の原子・分子に衝突して光を放ちます。⑪これが**オーロラ**です。

問17　極光是怎麼形成的呢？

答 ①太陽的表面總是**持續地釋放出電漿**。②所謂的電漿，是由一群帶有**正與負**離子的粒子的 集合體。③由於它像風一般地壟罩著太陽系流動，因此被稱之為**太陽風**。④地球以北極附 近為S極、以南極附近為N極，形成宛如一個巨大的磁鐵，外面包覆著磁場。⑤地球磁場受到太陽 風的影響，**被大大地牽引**到太陽的**相反方向**。⑥太陽本身也具有磁場。⑦強力的太陽磁場，足以 覆蓋地球磁場。⑧磁場具有**排斥電漿的特性**存在。⑨因此，當太陽磁場**連接**到地球磁場的時候， 電漿便進入地球磁場，帶著**強大的能量、沿著地球的磁力線飛奔**往南北極附近。⑩這時，電漿與 高空大氣層當中的氧或氮的原子與分子撞擊，釋放出光的現象。⑪這就稱之為極光。

Q18　Does air have weight?

①In daily life we do not feel the weight of air, but, in
日常生活　　　　　　　　　　　　空氣的重量

fact, it has weight. ②1,000 liters of air at 1 atmosphere
1公升　　　　　　　　1單位氣壓

weighs about 1.2 kilograms.
重量　　　　　　1.2公斤

③The weight of air, 1.2 kilograms per 1,000 liters, is only a
每

thousandth the weight of water, so you may think it is very light.
…的1000分之1重

④However, you are on the ground, which is at the bottom of
…所在的地表　　　　　　　　　底部

a thick layer of air. ⑤Air pressure is put on you on the ground just
厚厚一層　　　　　空氣的壓力　加在…上

like the heavy water pressure that is put on you when you dive
水壓　　　　　　　　　　　　　　　潛入…

into the deep sea. ⑥To be exact, air weighs 1 kilogram per square
深海　　　　　正確地來說　　　　　　　　　　每1平方公分

centimeter and 10 tons per square meter. ⑦To put it simply, air
每1平方公尺　　　　簡單地來說

weighs as much as the weight of 10 compact cars per square meter.
自小客車

⑧It is not easy to weigh air on the earth. ⑨If you put 1,000 liters
測量…的重量　　　　　　　　　裝入…

of air at 1 atmosphere in a bag and weigh it, it just weighs

0 grams. ⑩Since the weight of the air in
0公斤

the bag is the same as the buoyancy of
浮力

the air around it, scales say it is 0 grams.
測量

問18　空気に重さはあるの？

答 ①私たちの**日常生活**の中で、**空気の重さを感じることはありませんが、空気にも
重さがあります。②1気圧の空気の重さは1000リットルで約1.2キログラム**です。

③1000リットルで1.2キログラムという空気の重さは水の**1000分の1**であり、大した

ことはないと思うかもしれません。④しかし、**私たちがいる地上**は**厚い空気の層**であ
る大気の**底**です。⑤**深海に潜る**とすごい**水圧**がかかるように、地上にも**空気の圧力**が

かかっています。⑥それがどのくらいかというと、**1平方センチ**あたり1キロ、**1平方**

メートルで10トン。⑦つまり、1平方メートルあたり**小型乗用車**10台分にも相当する

のです。

⑧地球上で空気**の重さを量ろう**としても簡単ではありません。⑨1気圧の空気、1000リ

ットルを袋に**入れて**重さを量っても、**0グラム**にしかなりません。⑩袋の中の空気の重

さと周囲の空気の**浮力**とつり合ってしまうため、重さがないことになってしまうから
です。

問18　空氣有重量嗎？

答 ①雖然我們在**日常生活**當中，感受不到**空氣的重量**，但是空氣是有重量的。②在1單位氣壓
下，**1公升的空氣約重1.2公斤**。③每1公升重1.2公斤的空氣重量只有水的**1000分之1**，因此
或許你也幾乎感受不到重量。④然而**我們所處在的地表**，是位於厚厚一層空氣的大氣層底部。⑤
如同**潛入深海裡**感受到強力**水壓**一般，地面上也含有空氣的**壓力**。⑥那麼是多少比例呢?每1**平方
公分有1公斤、每1平方公尺就有10頓**。⑦**也就是說**，1平方公尺的空氣相當於**10台自小客車**的重
量。⑧在地球上**測量**空氣的重量並不是一件簡單的事。⑨假如將1單位氣壓的空氣**裝**1公升進入袋
子裡，會得到0公斤的重量。⑩那是由於袋子裡的空氣重量，與周圍的空氣**浮力**相同，因此無法
測量出重量。

The sky looks blue and sunset looks red because light is scattered.

空の青も夕焼けの赤も、光の散乱による
藍色的天空與紅色的夕陽，都是由於光的散色作用而形成

Blue sky
青空
藍天

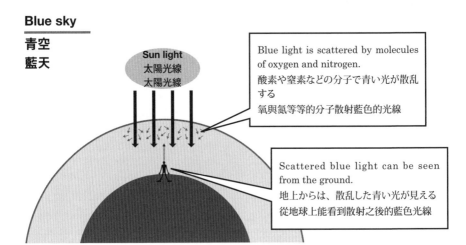

Sun light
太陽光線
太陽光線

Blue light is scattered by molecules of oxygen and nitrogen.
酸素や窒素などの分子で青い光が散乱する
氧與氮等等的分子散射藍色的光線

Scattered blue light can be seen from the ground.
地上からは、散乱した青い光が見える
從地球上能看到散射之後的藍色光線

Sunset
夕焼け
夕陽

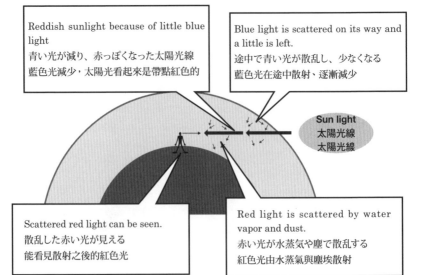

Reddish sunlight because of little blue light
青い光が減り、赤っぽくなった太陽光線
藍色光減少，太陽光看起來是帶點紅色的

Blue light is scattered on its way and a little is left.
途中で青い光が散乱し、少なくなる
藍色光在途中散射、逐漸減少

Sun light
太陽光線
太陽光線

Scattered red light can be seen.
散乱した赤い光が見える
能看見散射之後的紅色光

Red light is scattered by water vapor and dust.
赤い光が水蒸気や塵で散乱する
紅色光由水蒸氣與塵埃散射

Auroras are formed by light-emitting phenomena with plasma pulled by the earth's geomagnetism.

オーロラとは、地球の地磁気に引き込まれた プラズマによる発光現象のこと
極光的形成，是因為地球的磁場吸引電漿所引起的發光現象

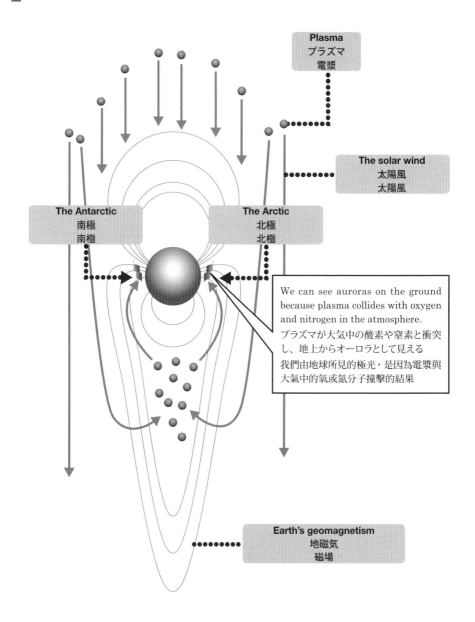

Plasma
プラズマ
電漿

The solar wind
太陽風
太陽風

The Antarctic
南極
南極

The Arctic
北極
北極

We can see auroras on the ground because plasma collides with oxygen and nitrogen in the atmosphere.
プラズマが大気中の酸素や窒素と衝突し、地上からオーロラとして見える
我們由地球所見的極光，是因為電漿與大氣中的氧或氮分子撞擊的結果

Earth's geomagnetism
地磁気
磁場

Q 19   How do earthquakes happen?

① The central part of the earth is called the core and the
(…的核心部分) (地核)

part around it is called the mantle. ② There is a part
(它的周圍部分) (地幔)

a few tens of kilometers thick called the crust outside the mantle,
(數十公里厚的…) (地殼)

which covers the surface of the earth.
(…的表面)

③ The mantle is a part made of layer of rocks from the inside of the
(由一層岩石組成) (由內側的…算起)

crust to a depth of 2,900 kilometers deep. ④ The outermost part of
(深達2900公里) (最外側的部分)

the mantle moves very little, but it is thought that the part from
(深度100～400公里的部分)

100 to 400 kilometers deep of it moves slowly as a result of heat
(由於…的熱對流)

convection. ⑤ In other words, the outermost part of the mantle and
(換言之)

the crust are floating together on the part which is moving
(在…上面一起漂浮)

on account of convection.
(因為…而導致)

⑥ These floating parts are called plate. ⑦ There are two kinds of
(板塊)

plate; continental plates and oceanic plates. ⑧ The surface of the
(大陸板塊) (海洋板塊)

earth is covered with more than ten plates. ⑨ Scientists think that

these plates are moved by the mantle's convection, and that they
(地幔的對流)

run into or overlap one another. ⑩ When some plates move,
(與…衝突) (重疊…) (互相)

a heavily loaded part may give way, cracking or sliding. ⑪ This is
(被施加強大的力量的部分) (無法負荷) (斷裂) (脫離)

how earthquakes happen.(see p.66)

問19　地震はどうして起きるの？

答 ①地球**の中心**にはコア（核）と呼ばれる部分があり、**その外側**にマントルがあります。②さらにその外側には**数十キロメートルの厚さの地殻**があり、これが地球の**表面**を覆っています。

③マントルは、地殻の下から深さ2900キロメートルまで続く岩石の層です。④**最上部**はほとんど動きませんが、**深さ100～400キロメートルの部分**は、**熱によりゆっくりと対流している**と考えられています。⑤つまり、対流によって動く層の上に、マントルの最上部と地殻がかたまりとなり、**浮いている**ような状態です。

⑥このかたまりは**プレート**と呼ばれています。⑦プレートには**大陸プレート**と**海洋プレート**があります。⑧地球の表面は10数枚のプレートで覆われています。⑨これらはマントルの対流により移動し、**ぶつかったり重なり合ったり**していると考えられています。⑩そして、プレートが移動して、ある部分に**大きな力が加わり、耐えられなくなると、そこがひび割れたり、ずれたり**します。⑪これが地震です。

問19　地震是如何引起的呢？

答 ①地球的中心有個被稱之為**地核**的核心部分，其外側包覆著一層**地幔**。②緊接著地幔有著一層數十公里厚的地殼，覆蓋在地球的表面。③地幔是由地殼下方起算，深達2900公里厚的一層岩石組成。④雖然最上面的部分幾乎不會移動，但根據研究，深度100～400公里的地函部分，由於地熱的緣故會慢慢地對流。⑤換言之，位於對流層之上，地幔的最上層與地殼連接的部分是處於漂浮的狀態。⑥這個漂浮的部分我們稱之為**板塊**。⑦板塊之中又分為**大陸板塊**與**海洋板塊**。⑧地球的表面共由10數枚的板塊包覆著。⑨科學家認為，這些板塊會因**地幔的對流**而移動，並且互相撞擊或彼此重疊。⑩所以當板塊移動的時候，某個部分會被施加強大的力量，當無法負荷時就會斷裂或者脫離。⑪這就會發生地震。

Q20 How were the deserts formed?

①You may think of the deserts as sand areas from their
可以想像…　　　　　　　　　　　　沙地

Japanese name, sabaku. ②However, sand areas are

called sandy deserts. There are also many rock deserts in the
砂質沙漠　　　　　　　　　　　　　　岩質沙漠

world, in which you can see a lot of rocks.
岩石

③Deserts are formed in environments in which a lot of water
因…而形成　藉由環境

evaporates rather than in environment in which it rains a lot. ④
蒸發　　　比起…還多

When plants are not likely to grow because of dry air or salt in the
岩石　　不容易…　　　　　　　　　空氣乾燥　　鹽分

ground, the topsoil which has organic matters is washed away or
表土　　　　　　有機物質　　　被沖刷

blown away. ⑤This makes a dead land in which it is difficult for
被風吹散　　　　　　　　不毛的

animals and plants to live. ⑥This is how the deserts are formed.
如何

⑦The characteristic of a desert climate is the big difference in the
特徵　　　　　　沙漠的氣候　　　　　　由於…之間的氣溫差異

temperature between day and night because it is dry. ⑧The

surface temperature of rocks goes up in direct sunlight during the
表面溫度　　　　　　　上升　陽光直接照射　白天

day and it goes down at night. ⑨In the course of this change of
下降　　　　　　　在…的過程中

temperature, broken-up rocks are blown away or hit each other
崩解　　　　　　　　　　　彼此撞擊

and become small pieces. ⑩The sand made in this way is piled up
藉由這個方式　逐漸堆積形成了…

to make a sandy desert.

問20　砂漠はどうやってできたの？

答 ①「砂漠（さばく）」というと、**砂地**だと思いがちです。②しかし、砂に覆われているのは**砂砂漠（すなさばく）**であり、世界的に見て多いのは岩肌がむき出しになった**岩石砂漠**です。

③**砂漠ができる原因**となるのは、降水量よりも水の**蒸発量**が多い**環境**です。④**乾燥**や地表の**塩分**などにより**植物**が**育ち**にくくなると、**有機物**を含んだ**表土**が**流され**たり風に飛ばされたりしてなくなります。⑤その**結果**、さらに動植物が生息しにくい**不毛の地**と化してしまいます。⑥これが砂漠です。

⑦**乾燥**しているため、**昼夜の気温の差**が大きいことが**砂漠の気候の特徴**です。⑧**昼間**は**直射日光を浴びた**岩の**表面温度は上昇**し、夜になると冷えます。⑨これを**繰り返して**いるうちに、岩はぼろぼろに**崩れ**、風に飛ばされたり、**ぶつかり合ったり**して細かくなります。⑩**こうしてできた砂**が**堆積**すると、砂砂漠になります。

問20　沙漠是如何形成的呢？

答 ①所謂的「沙漠」，可以想成是由沙子組成的地面。②但是，由沙子所覆蓋的稱為**砂質沙漠**，在世界上較常見的還有由岩石所覆蓋的**岩質沙漠**。③會形成沙漠的原因，是由於**環境**中的水分蒸發量比起降雨量還多的緣故。④由於空氣**乾燥**與地表含有**鹽分**，導致**植物生長困難**，含有有機物質的表土容易被沖刷或被風吹散。⑤這樣的結果，形成動植物都難以生長的**不毛之地**。⑥這就是沙漠。

⑦由於乾燥的緣故，日夜**溫差**非常大為**沙漠氣候的特徵**。⑧白天由於**陽光直接照射**，岩石表面的**溫度上升**，到了夜晚便逐漸冷卻。⑨在這樣溫度反覆地變化之中，岩石逐漸**崩解**、被風吹拂撞擊，顆粒漸漸地變細。⑩**變成沙粒之後、逐漸堆積**，便形成了砂質沙漠。

Q21 Which is colder, the North Pole or the South Pole?

①You may think the North Pole and the South Pole are of the same temperature because they are both polar areas, but, in fact, the South Pole is colder.

②The South Pole is a continent and has various landform changes, as is the case with other continents. ③There are mountains such as the Ellsworth Mountains and the Yamato Mountains. ④The highest mountain is Vinson Massif, which is 4,892 meters above sea level and higher than the Matterhorn in the Alps. ⑤The North Pole is a sea covered with ice. ⑥The ice is about ten meters thick on average and there are no high places unlike the South Pole.

⑦In general, when you compare temperatures in places on the same latitude, land is likely to be colder than sea and highlands are likely to be colder than lowlands. ⑧The reason why the South Pole is colder is that it is land and is higher above sea level. ⑨It is said that the average temperature at the South Pole is about 20 degrees C lower than that at the North Pole. ⑩The lowest temperature measured on earth is minus 89.2 degrees C, recorded at Vostok Station at the South Pole on July 21, 1983.

問21　北極と南極では、どちらが寒いの？

答 ①<u>北極</u>と<u>南極</u>はどちらも<u>極地</u>であるため、気温も変らないように思う**かもしれま**
<small>北極　南極　　　　極地</small>

せんが、寒いのは南極のほうです。

②<u>大陸</u>である<u>南極</u>は、ほかの大陸と同じように<u>地形の変化</u>に富んでいます。③**エルスワ**
<small>大陸　南極　　　　　　　　地形　変化</small>

ース<u>山脈</u>、やまと山脈などの山脈があります。④<u>最高峰</u>である**ヴィンソン・マシフ山**
<small>山脈　　　　　　　　　　　最高峰</small>

は標高4,892メートルと、アルプスのマッターホルンをしのぎます。
<small>標高</small>

⑤北極は、<u>氷</u>に覆われた海です。⑥氷の厚さは<u>平均</u>10メートルぐらいで、南極のよう
<small>氷塊　　　　　　　　平均</small>

に標高が高い場所がありません。

⑦<u>一般的</u>には、<u>同緯度</u>の場所を<u>比較</u>すると、海よりも陸、低地よりも高地の気温が低
<small>　　　同緯度　　　比較</small>

くなる傾向があります。⑧南極が寒いのも、陸地であり標高が高いという<u>理由</u>からで
<small>　　　　　　　　　　　　　　　　　　　　　　因素</small>

す。⑨<u>平均気温</u>を比較すると、南極のほうが北極より20度ほど低いとされています。
<small>平均氣温</small>

⑩また、地球上で<u>観測</u>された<u>最低気温</u>は、1983年7月21日に南極の**ボストーク基地**で
<small>　　　　観測　最低温度　　　　　　　　　　　　　　站</small>

<u>記録</u>されたマイナス89.2度です。
<small>記録</small>

問21　北極或南極，哪一個地方比較寒冷呢？

答 ①雖然北極與南極兩個地區都位於極地的位置，你也許會認為他們的氣溫是相同的，但實際上南極是比較寒冷的。②南極是一個大陸，與其他的大陸相同有豐富的地形變化。③這裡有艾爾史渥斯山脈與大和山脈等等的山脈。④身為最高峰的文森峰，標高4,892公尺，甚至凌駕於阿爾卑斯山脈的馬特杭峰。⑤北極是一片覆蓋著冰塊的海洋。⑥冰塊的厚度平均約10公尺，沒有像南極一樣的高地存在。⑦一般來說，比較同緯度地區的溫度時，陸地比起海洋、高地比起低地的氣溫會較低。⑧南極之所以比較冷，是因為身為陸地且標高比較高的因素。⑨平均氣溫相較，南極的氣溫低於北極約攝氏20度。⑩此外，從地球上所觀測到的最低溫度，是在1983年7月21日位於南極的沃斯托克站所記錄的攝氏負89.2度。

Q 22 Why doesn't lake water soak into the bottom?

①The bottom of a lake is usually covered with pebbles,
…的底部　　　　　　　　　　覆蓋著…　　　小石礫

sand, and so on. ②However, a lake is always full of
　　等等　　　　　　　　　　　　　　充滿著…

water. ③So, why doesn't lake water soak into the bottom of a lake?
　　　那麼　　　　　　　　　　滲入…

④Since the bottom of a lake is covered with pebbles, sand, soil,
因為…　　　　　　　　　　　　　　　　　　　　土

and so on, water actually soaks into the bottom. ⑤However, there
　　　　　　　實際上

are always layers of rock or clay under the bottom and these layers
　　　　　岩石或粘土層

do not take in much water.
　　讓…進入

⑥The water which soaks into the pebbles and sand piles up on
　　　　　　　　　　　　　　　　　　　　　聚集在…

these layers. ⑦After that, some of the water flows away as
　　　　　　　　　　　　　　　　　　　流出來

underground water.
作為地下水

⑧It is true that some water soaks into the bottom of a lake or
正確地來說…

evaporates from the surface of the lake, but it seems that lake
由…蒸發　　　　…的表面

water does not soak into the bottom of

a lake because the same amount of
　　　　　　　等量的水

water is added by rivers, rain, melted
由…補充　　　　　　　融化的

snow, and so on. (see p.67)

問22　湖の水はどうして、底にしみ込まないの？

答　①湖の底は多くの場合、**小石**や砂などで**覆われています**。②にもかかわらず、湖
　　　　地方　小石礫

はいつも水をたたえています。③なぜ、湖の水は湖底にしみ込まないのでしょうか？
　　　　　　　　　　　　　　　　　　　湖底

④湖の底は、小石や砂、土などで覆われており、水がしみ込みます。⑤しかし、その下

には必ず**岩や粘土の層**があり、ほとんど水を通しません。⑥湖の底の砂や小石にしみ
　　　　　　粘土

込んだ水は、この層の上に**たまります**。⑦そして、その一部は**地下水となって流出し**
　　　　　　　　　　　　　　　　　　　　　　　一部分

ます。

⑧湖底にしみ込んだり、湖面**から蒸発する**水もありますが、流れ込む川や、雨、雪ど
　　　　　　　　　　湖面　　　蒸發

けによって**水が補給される**ために、湖の底に水がしみ込んでいないように見えるので
　　　　　　　補給

す。

問22　湖水為什麼不會滲入地底？

答　①湖泊的底部，大部分都覆蓋的小石礫、沙粒等等物質。②然而，湖泊卻總是**充滿著水**。
③為什麼湖水**不會滲入湖底**呢？④湖泊的底部，覆蓋著小石礫、沙、土等等物質，實際上
會滲入地底。⑤但是，在這一層的下面還覆蓋著岩石與粘土層，湖水幾乎不會滲入。⑥滲入湖底
的沙粒、石礫這一層的湖水，會聚積在粘土層的上面。⑦接著，其中一部分會**成為地下水流出**。
⑧事實上湖水會滲入地底、或經由湖面蒸發，但是因為河川的水流匯集，以及雨水、雪水的**水量
補給**，使得湖水看起來像是不會滲入地底。

Q23 Why are snow crystals hexagonal?

①There are various shapes of snow crystals but perfectly
各式各樣的形狀　　雪的結晶　　　完美的成形

formed ones are all hexagonal. ②If you look over, you
六角形的　　　　　調查

will find that not only snow crystals but also ice crystals such as
不僅…而且…　　　　　　冰塊的結晶　例如

frost are hexagonal.
霜

③A water molecule is made up of two hydrogen atoms and one
水分子　　　　　由…構成　　氫原子

oxygen atom. ④Two hydrogen atoms form an angle of 105 degrees.
氧原子　　　　　　　　　　排列成…　…的角度　105度

⑤When water molecules get together and other water molecules
聚集

attach to empty spaces, three hydrogen atoms attached to one
連結　　空曠的地方

oxygen atom in three directions form an angle of 120 degrees. ⑥
從3個方向　　以120度的角度排列而成

That is the group of water molecules which is in the state of
安定的狀態

stability.

⑦Snow is Water crystallized in the form of thin plates in air. ⑧
結晶　　薄板狀

What is needed most in forming snow crystals is the angle of 120
基本的條件

degrees of hydrogen atoms. ⑨You can find snow crystals in

various shapes, but they are all hexagonnal because what is
各式各樣的形狀

needed most is the angle of 120 degrees of

hydrogen atoms. (see p.67)

問23　雪の結晶はなぜ、六角形なの？

答 ①雪の結晶にはさまざまな形のものがありますが、美しく成長したものはすべて呈

六角形です。②雪だけでなく、霜柱など氷の結晶を調べてみると、やはり六角形をし

ていることがわかります。

③水の分子は2個の水素原子と1個の酸素原子が結びついています。④2個の水素は、

105度の角度を保っています。⑤そして、水の分子が集まると、空いている場所にほか

の水分子がくっつき、120度の間隔で3方向に水素がついている形になります。⑥これ

が水分子が集まって安定した状態です。

⑦雪は、水が空気中で薄い板状に結晶化したものです。⑧雪の結晶ができるときに基本

となるのは、水素同士の120度という角度です。⑨雪の結晶にはいろいろな形がありま

すが、どの形も水素原子の120度が基本となるため、六角形になるのです。

問23　雪的結晶為什麼呈六角形？

答 ①雖然雪的結晶有各式各樣的形狀，但是完美的結晶是呈美麗的六角形。②不只是雪，仔
細地觀察霜與冰塊的結晶，也是呈現六角形。③水分子是由2個氫原子與1個氧原子結合而
成。④2個氫原子原本保持著105度的角度。⑤然而，當水分子聚集時，在空曠的地方與其他的水
分子連結，是以120度的間隔從3個方向集合成氫原子的形狀。⑥這就是水分子呈安定狀態的聚
集。

⑦雪是水在空氣中形成的薄板狀的結晶。⑧要形成雪的結晶時，基本的條件必須要水分子呈120度
角。⑨雖然雪的結晶有各式各樣的形狀，但不管哪一種形狀的水原子都呈現基本的120度角，因
此就形成了六角形。

Most earthquakes happen when plates in the earth run into one another.

地震の多くは、地球のプレートが
ぶつかり合って起きる
大多數的地震發生，是由於地球板塊的撞擊所引起的

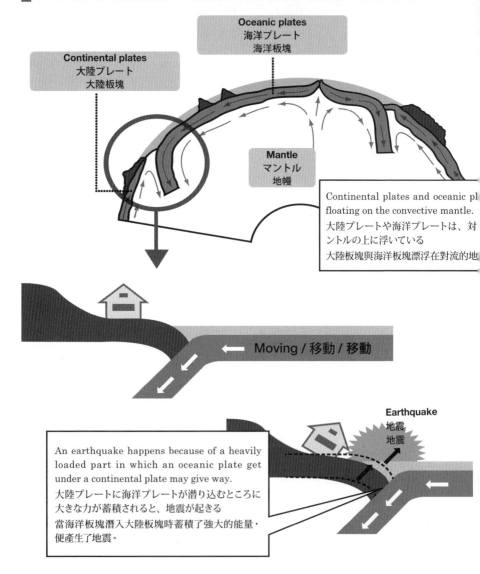

Continental plates
大陸プレート
大陸板塊

Oceanic plates
海洋プレート
海洋板塊

Mantle
マントル
地幔

Continental plates and oceanic pl
floating on the convective mantle.
大陸プレートや海洋プレートは、対
ントルの上に浮いている
大陸板塊與海洋板塊漂浮在對流的地

Moving / 移動 / 移動

An earthquake happens because of a heavily
loaded part in which an oceanic plate get
under a continental plate may give way.
大陸プレートに海洋プレートが潜り込むところに
大きな力が蓄積されると、地震が起きる
當海洋板塊潛入大陸板塊時蓄積了強大的能量，
便產生了地震。

Earthquake
地震
地震

A lake is always full of water because there are layers under the bottom and these layers do not take in water.

湖の水がなくならないのは、
湖底に水を通さない層があるから
湖水不會流失的原因是因為，
湖底有一層水無法通過的粘土層

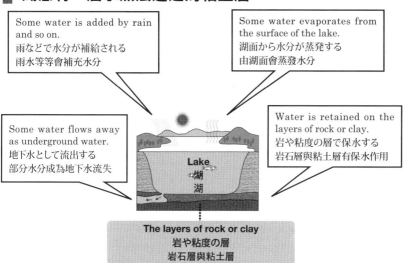

Some water is added by rain and so on.
雨などで水分が補給される
雨水等等會補充水分

Some water evaporates from the surface of the lake.
湖面から水分が蒸発する
由湖面會蒸發水分

Some water flows away as underground water.
地下水として流出する
部分水分成為地下水流失

Water is retained on the layers of rock or clay.
岩や粘度の層で保水する
岩石層與粘土層有保水作用

Lake
湖
湖

The layers of rock or clay
岩や粘度の層
岩石層與粘土層

Snow crystals are combination of hydrogen bonds and they are hexagonal.

雪の結晶の形は、水素結合の
組み合わせで六角形になる
雪結晶的形狀，是由氫原子相互結合之後
形成了六角形的形狀

Hydrogen bond
水素結合
氫原子結合

H 120°
120° O H O
H 120°
H

Hydrogen bond, in which three hydrogen atoms attached to one oxygen atom in three directions form an angle of 120 degrees is in the most stable state.
酸素に対して、120度の間隔で3方向に水素がついた水素結合は、水分子がもっとも安定した状態
氫原子的結合是由3個方向的氫原子以120度角的間隔與氧原子結合，形成最安定的水分子狀態。

Q24 Why does it snow though the temperature is not below 0 degrees C?

① Snow is formed when water vapor frozen high in the air becomes ice particles which turn into big crystals as they fall. ② It snows when it is below 0 degrees C up in the air regardless of the temperature on the ground. ③ You may think, "Why does not snow melt into rain as it falls when the temperature on the ground is above 0 degrees C?"

④ Even when it is warmer than 0 degrees C, it sometimes snows without the snow melting. ⑤ This is because snow needs heat around it when it melts. ⑥ Heat is needed to raise the temperature of snow to 0 degrees C, the melting point of ice. ⑦ It needs more heat to melt snow. ⑧ It needs as much as about 80 kilocalories of heat per kilogram when ice melts and turns into water. ⑨ This amount of heat can raise the same amount of water from 0 degrees C to 80 degrees C.

⑩ That is why snow stays on the ground without melting even when you feel warm. ⑪ For the same reason, snow sometimes falls on the ground without melting in the air even when the temperature is above 0 degrees C.

問24　気温が零度以下でなくても、なぜ雪が降るの？

答 ①雪は、**上空**で**水蒸気が冷**やされてできた**氷の粒**が、落下中に成長して大きな結
（天空）（水蒸氣）　　　　　　　　　　　　　　　（降落）　（變成）

晶となったものです。②**地上の気温に関係なく、上空が氷点下**なら雪になります。
（地面）　　（關係）　　　（天空）（冰點以下）

③「地上の気温が**0℃以上**だったら、雪は空中で**融けて雨になる**のでは？」と思うでし
　　　　　　　　　　　　　　　　　　と

ょう。

④気温が0℃より高くても、**雪が融けず**に降ることがあります。⑤**これは、雪が融ける**
（氣溫）

とき、周囲の**熱**が必要だからです。⑥まず雪の温度を**融点**の0℃まで**上昇させる**ために
（周圍）（必須）　　　　　　　　　　（溫度）（融點）　ゆうてん　（提升）

熱が必要です。⑦雪が融けるには、さらに熱が必要です。⑧氷が融けて水になるために

は、**1キログラムあたり約80キロカロリー**もの熱が必要なのです。⑨これは**同量の水**
　　　　　　　　　　　　　　　　　　　　　　　　　　　　　　　　　（等量）

を0℃から80℃まで温めることができるほどの熱量です。⑩ポカポカと暖かく感じて
　　　　　　　　　　　　　　　　　　　（熱量）

も、積もった雪が融けず**に残るのは、このような理由**からです。⑪気温が0℃以上で
　　　　　　　　　　　　　　　　　　　　（原因）

も、雪が空中で融けることなく**地上に達する**ことがあるのも、**同じ理由です。**
　　　（空中）

問24　氣溫不到零度以下，為什麼會下雪呢？

答 ①雪是由天空中的水蒸氣冷卻後所形成的冰塊，在降落的過程中變成比較大的結晶的物質。②**與地面上的氣溫無關**，是因為天空低於冰點以下而形成了雪。

③那麼你是不是會有疑問「地面上的氣溫要是高於0℃以上，雪在空中會融化而形成雨降落嗎？」

④即使氣溫高於0℃，**雪還是不會融化而照常降落。**⑤**這是因為**雪在融化的時候，周圍必須要有熱度。⑥首先，必須要有能夠將雪的溫度提升到0℃以上的融點的熱度。⑦要讓雪融化，還必須要持續的熱度。⑧要讓冰融化成水，**每公斤需要消耗80大卡的熱量。**⑨這相當於將等量的水約由0℃加熱到80℃的熱量。⑩這也就是為什麼即使我們感覺很溫暖，地面上還是有積雪不會融化的原因。⑪即使氣溫到達了0℃以上，雪在空中並不會融化而**降落到地面，**其原因也相同。

Let's see the world map.

世界地図をみてみよう
讓我們來看看世界地圖吧

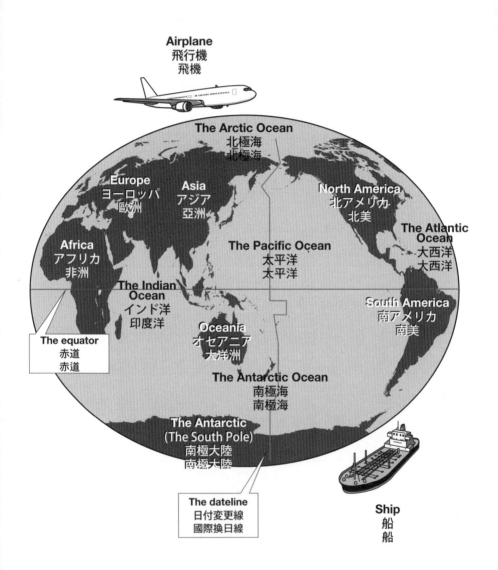

Airplane
飛行機
飛機

The Arctic Ocean
北極海
北極海

Europe
ヨーロッパ
歐洲

Asia
アジア
亞洲

North America
北アメリカ
北美

The Atlantic Ocean
大西洋
大西洋

Africa
アフリカ
非洲

The Pacific Ocean
太平洋
太平洋

The Indian Ocean
インド洋
印度洋

South America
南アメリカ
南美

The equator
赤道
赤道

Oceania
オセアニア
大洋洲

The Antarctic Ocean
南極海
南極海

The Antarctic
(The South Pole)
南極大陸
南極大陸

The dateline
日付変更線
國際換日線

Ship
船
船

Chapter 3

Questions about creatures

第 3 章

生き物の疑問
生物的疑問

① It is said that dinosaurs, which ruled the world for
about 150 million years from over 200 million years ago,
died out 65 million years ago at the end of the Mesozoic era in
the Cretaceous period. ② There are various unproved theories
about the reason they died out such as an epidemic, a volcanic
eruption, and so on. ③ Among them, the most believable theory is
a meteor strike.

④ The basis for this theory is a thin layer of metal, iridium, which
covers the layer of the time when dinosaurs are thought to have
died out. ⑤ The iridium layer is found all over the world.
⑥ Since there is almost no iridium on earth, it is thought that this
metal was brought to earth by a meteor or something like that.
⑦ Scientists say that a huge meteor, which struck the earth at the
end of the Cretaceous period and made a crater of about 180
kilometers across on the Yucatan Peninsula, formed this layer.
⑧ This huge meteor is about 10 kilometers across and, when it
struck the earth, it might have given off energy about 5 billion
times as much as the atomic bomb dropped over Hiroshima. ⑨ The
most believable theory says that, because of this strike, little
sunlight reached the surface of the earth on account of the great
deal of dust with iridium which rose up, causing the whole earth
to get colder and dinosaurs to die out.

問25　恐竜はなぜ、絶滅したの？

答 ①2億年以上前から約1億5000万年間、地球を支配したとされる恐竜が絶滅したのは、今か
ら約6500万年前にあたる**中生代白亜紀**の末期だといわれています。②その理由については、
伝染病説、火山の爆発などさまざまな説があります。③そのなかでも**有力視されているの**
が、隕石衝突説です。

④その**根拠**となるのは、恐竜が絶滅したと思われる時代の地層を覆う、**イリジウムという金属の**
薄い層です。⑤これは世界各地で発見されています。⑥イリジウムは、地球にはほとんど存在し
ないため、隕石などによって宇宙から**もたらされた**と考えられます。⑦イリジウムの層をつくった
のは、白亜紀末期に地球に**衝突し、ユカタン半島に直径約180キロメートルのクレーター**をつく
った**巨大隕石**だと考えられています。

⑧この巨大隕石は直径約10キロメートルほどもあり、衝突時のエネルギーは、広島型**原爆**の約
50億倍と推定されています。⑨衝突により、イリジウムを含む**大量の粉塵が舞い上がって地表に**
届く太陽光線が減少した結果、地球全体が寒冷化し、恐竜が絶滅したという説が有力です。

問25　恐龍為什麼會滅絕？

答 ①據說，曾經在2億年以前主宰地球約1億5千萬年的恐龍，滅絕於距今6500萬年前的**中生代**
白堊紀末期。②關於其滅絕的理由，有傳染病說、火山爆發說等等各式各樣的推測。③其
中最有說服力的，莫過於隕石撞擊說。④這個理論的**根據**在於，**據信**是恐龍滅絕時代的地層，所
覆蓋的一層被稱之為銥的金屬地層。⑤這一層金屬在世界各地都有被發現。⑥由於銥金屬幾乎不
存在地球之上，據信是由隕石等等媒介由宇宙帶往地球。⑦科學家認為，巨大隕石在白堊紀末期
撞擊地球的尤加敦半島，造成**直徑**約180公里的隕石坑，並且形成了銥金屬層。⑧這顆巨大隕石
直徑約10公里，在撞擊時發出的能量，據推算約為廣島**核爆**的50億倍。⑨這個最有力的學說認
為，因為這個撞擊，含有大量銥金屬的粉塵飄到地表之上、並阻止了太陽光線照射，整個地球趨
於寒冷、造成恐龍滅絕。

Q26 Why do animals have two sexes, male and female?

A ① Some species have one sex but can reproduce themselves. ② One example of this is potatoes in the ground. ③ Potato plants flowers and produce fruits but the potato is a part of a tuber changed in shape and it is produced without pollination. ④ If you put a potato in the ground, it will put out buds and its "child" grows up. ⑤ A "child" reproduced without pollination in this way is a clone which has the same genes as its "parent". ⑥ So if the "parent" is sensitive to the cold, the "child" is also sensitive to the cold.

⑦ Mammals such as human beings reproduce themselves with fertilization, in which the female's egg nucleus and the male's sperm nucleus unite, and the child gets two halves of the genes which each of the parents has. ⑧ In this case, even if the child's father is sensitive to the cold, the child is not always sensitive to the cold. ⑨ Mammals can have various children because of the combination of the male's and the female's genes. ⑩ A child of parents who are likely to be infected by a certain virus is sometimes not likely to be infected by the virus.

74

問26　動物はなぜ、雄と雌に分かれているの？

答 ①生物の中には、雄と雌がいなくても子孫を増やせる種もあります。②たとえば、土の中にできるジャガイモです。③ジャガイモには花が咲き実もできますが、ジャガイモのいもは茎の一部が変形してできたもので、受精とはまったく関係なくできます。④これを土に埋めておけば、芽が出て「子」が育ちます。⑤このように、受精なしにできた子は、親とまったく同じ遺伝子を持つクローンです。⑥そこで、たどえば親が寒さに弱い個体だとすると、子も寒さに弱くなります。

⑦人間をはじめとする哺乳類のように、雌の卵の核と雄の精子の核とが合体する受精によって子ができる場合には、子は両親の遺伝子を半分ずつ受け継ぎます。⑧この場合、父親が寒さに弱いとしても、母親の遺伝子も受け継いでいる子は寒さに弱くなるとは限りません。⑨雄と雌の遺伝子の組み合わせにより多様な子孫を残せるのです。⑩同様に、特定のウィルスに感染しやすい個体の子孫が、同じウィルスに感染しにくくなることもあります。

問26　動物為什麼有雌雄的分別？

答 ①在生物之中，也有不分雌雄仍然可以繁殖的物種。②舉例來說，生長在土中的馬鈴薯。③馬鈴薯會開花、結果實，而馬鈴薯的球根是由莖的一部份變化而來，與受精完全沒有關係。④若是將馬鈴薯埋入土中，就會發芽、並且培育「子孫」。⑤像這樣無須經過受精而繁衍出來的後代，與親代之間擁有相同遺傳因子稱之為無性生殖。⑥因此，假如親代屬於畏寒的個體，子代同樣也會畏寒。⑦身為哺乳類之首的人類，是經由雌性卵子的核與雄性精子的核結合受精而產生子代，子代分別繼承雙親的一半遺傳因子。⑧在這種情況下，雖然父親屬於畏寒的體質，但同樣接受母親遺傳因子的子代，並不一定畏寒。⑨因為雄性與雌性遺傳因子的組合，產生了各種不同的子代。⑩同樣的，容易感染特定病毒的個體的子孫，也有可能不易受到相同病毒的感染。

Q27 Why Can Sparrows perch on Electric Wires?

① You can sometimes see sparrows or crows perched on
麻雀　　　　　　　　　　　　烏鴉　　　　　棲息在…

the stripped overhead wires of the railroad. ② Why don't
　　　裸露的　　　電車所架設的電線

they get an electric shock?
　　　觸電

③ Getting an electric shock is when an electric current flows
　　　　　　　　　　　　　　　　　　　　　電流　　　　　　　　　流經…

through the body. ④ An electric current flows through the body

when each of the two points on the body touches two things which
　　　各自的…　　體內2點

have different voltages. ⑤ Birds perch on one electric wire with
　　　不同的電壓　　　　　　　　　　　　　電線　　　　　雙腳

both legs. ⑥ Even if a 1,000-volt current flows through this electric
　　　　　即使…　　1000伏特的電流

wire, they do not get an electric shock because both of their legs
　　　　　　　　　　　　　　　　　　　　　　　兩個…

touch the electronic wire.

⑦ You are in danger of getting an electric shock when you fly a kite
　　　發生…的危險　　　　　　　　　　　　　　　　　　放風箏

and it gets caught on the overhead wires of the railroad. ⑧ That's
　　　掛在…之上　　　　　　　　　　　　　　　　　　那是由於…

because the ground has 0 volts and the overhead wires of the
　　　　地面

railroad have 1,000 volts or something. ⑨ An electric current flows

through the kite line to you, and then, to the ground because these
流經風箏線

two points have different voltages. (see p.80)

問27　スズメはなぜ、電線にとまっても平気なの？

答　①スズメやカラスは平気で電線がむき出しになった電車の架線などにとまっていることがあります。②なぜ、感電しないのでしょうか。

③感電とは、体に電流が流れて衝撃を受けることをいいます。④体に電流が流れるのは、電圧に差がある二つのものを体の2点で触れたときです。⑤鳥は一本の電線に両足をそろえてとまります。⑥その電線がたとえ1000ボルトだとしても、両足ともに

1000ボルトに触れていることになり、感電しません。

⑦凧あげをしているとき、凧が電車の架線にひっかかった場合などは、感電する危険があります。⑧架線の電圧が1000ボルトだとして、地面は常に0ボルトです。⑨2点に電圧の差があるため、凧糸—人間—地面と電流が流れるからです。

問27　為什麼麻雀可以平安無事地停在電線上？

答　①你經常可以看見麻雀或是烏鴉，停留在電車所架設的裸露的電線上等等地方。②為什麼牠們不會觸電呢？

③所謂的觸電，是電流通過體內所產生的衝擊。④當電流流入體內，電壓不同的兩種電流在體內2點觸及的時候發生。⑤鳥類用雙腳停留在一根電線上。⑥假設那根電線為1000伏特，因為雙腳同時接觸1000伏特，並不會觸電。⑦在放風箏的時候，風箏若是掛在電車的電線上，就可能發生觸電的危險。⑧由於電線的電壓為1000伏特，地面通常為0伏特。⑨因為2處的電壓不同，電流經由風箏線流經人體到達地面而導致觸電。

Q28 Why do migratory birds fly in a V-formation?

①Migratory birds move their wings up and down when
候鳥　　　　　　　　　　翅膀　上下地
they fly. ②When they lower their wings, the air pressure
　　　　　　　　　　　向下　　　　　　　氣壓
under the wings becomes higher than that of above them and this
　　　　　　　　　　　　　　　　　翅膀上面的氣壓
provides lifting power. ③This lifting power make birds fly in the
給予　　升力
sky.

④When birds move their wings, the flow of air changes around the
　　　　　　　　　　　　　　　氣流
tip of the wings.⑤This is because some of the air pushed down by
…的先端　　　　　　這是因為　　　　　　　　　因…而往下壓
the wings goes up around the wings at both tips of the wings.
　　　　往…的上方回流
⑥This flow of air makes two vortexes behind each wing of the
　　　　　　　　　　　　　漩渦
flying bird. ⑦These are called wing-tip vortexes in aeronautics.
　　　　　　　　　　　　　翼尖漩渦　　　　　航空學
⑧Air goes up in some parts of this flow of air and goes down in
　　　　　　在某些…的部分
other parts of this flow of air making vortexes. ⑨A bird flying
在其他…的部分　　　　　　　　　產生漩渦
behind the bird making vortexes finds it easy to fly on the
在…的後面飛行　　　　　　　　　　　　　　　　乘著上升的氣流飛行
ascending air of the air flow.

⑩Therefore migratory birds fly in a V-formation because the birds
　　　　　　　　　　　　呈V字編隊飛行
are following the ascending air of the wing-tip vortexes made by
追隨著…的腳步
the bird flying ahead of them. (see p.81)
　　　　在…的前面飛行

78

問28 渡り鳥はどうして、Vの字になって飛ぶの？

答 ①**渡り鳥**が飛んでいるときには、**翼を上下**に動かしています。②**翼を下げた**と

き、翼の下の**気圧**が上の気圧よりも高くなり、**揚力**が発生します。③この揚力によっ

て、鳥は空を飛ぶことができるのです。

④鳥が翼を動かすと、**翼の先**のほうだけは**空気の流れ**が変わります。⑤翼の下に押し下

げられた空気が、**両翼の先端**で上に回り込んでしまうからです。⑥この空気の流れ

は、飛んでいる鳥の後方で両翼分の2本の**渦**となります。⑦**航空用語**でいう**翼端渦**で

す。

⑧この空気の流れは渦を巻いていますから、空気が**上昇**しているところと**下降**すると

ころがあります。⑨**後続**の鳥は、空気が**上昇**しているところを飛べば、楽に飛べるわ

けです。

⑩つまり、渡り鳥が**V字編隊**になるのは、すぐ前の鳥が残した翼端渦が上昇している

部分を追って後続する鳥が飛んでいるからです。

問28 候鳥為什麼呈V字型飛行？

答 ①候鳥在飛行的時候，**翅膀**會上下地拍動。②當**翅膀向下**時，翅膀下的氣壓高於上面的氣壓而產生**升力**。③藉由這個升力，鳥類可以在空中飛行。

④鳥類在拍動翅膀的時候，只有翅膀**先端**的氣流會變化。⑤那是由於**拍入翅膀下方**的空氣，經由兩翼的先端**回流至上方**的緣故。⑥這個空氣的流動，會在飛行中的鳥類後方，形成分成兩翼的條狀漩渦。⑦**航空學**上稱做為**翼尖漩渦**。⑧在空氣上升時與空氣下降時，由於空氣的流動而產生漩渦。⑨後續的飛鳥若是在**空氣上升的時候飛行**，就能輕鬆地搭上上升氣流飛行。⑩也就是說，候鳥為什麼會呈**V字編隊**飛行，是由於追隨最前頭飛鳥的翼尖漩渦產生的上升氣流所致。

Birds do not get an electric shock if they touch, with both legs, two points which have same voltage.

両足の間に電圧差がなければ、鳥は感電しない
兩腳之間的電壓若是沒有差距，鳥類不會觸電

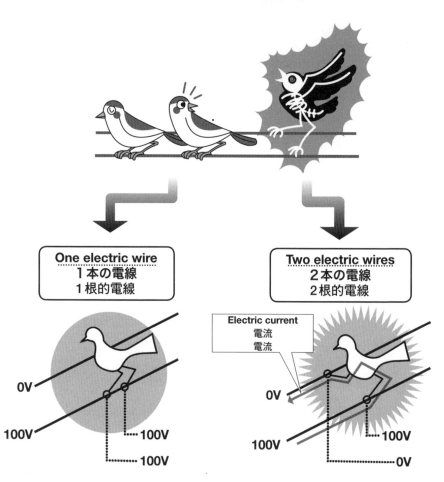

One electric wire
1本の電線
1根的電線

Two electric wires
2本の電線
2根的電線

Electric current
電流
電流

0V

100V ·····100V

·····100V

0V

100V ·····100V

·····0V

Its both legs touch the same voltage, 100-volt, so no electric current flows through its body.

両足ともに100ボルトであり、電圧に差がないため、電流が流れない

兩腳同時接觸100伏特的電壓，由於電壓沒有差距，並不會產生電流。

One of its legs touches 0-volt and the other touches 100-volt, in which voltage difference are 100-volt, so a large electric current flows through its body.

片足が0V、もう一方の足が100Vなら、両足に100Vの電圧差があるため、大き電流が流れる

假如一隻腳接觸0V，另一隻腳接觸100V，由於兩隻腳有100V的電壓差，因此大量的電流會開始流動。

Migratory birds fly in a V-formation because a bird flies on the vortex made by the bird flying ahead of it.

渡り鳥は、前の鳥が作った
空気の渦の上を飛ぶのでV字になる

候鳥在飛行時，由於後面的鳥乘著領頭鳥所製造的翼尖漩
渦而飛行，因此呈V字型

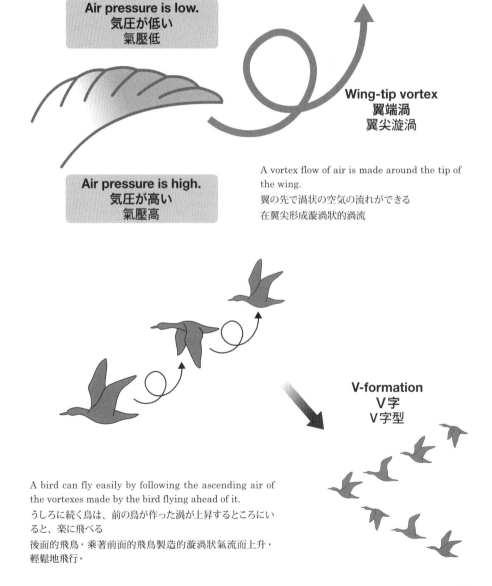

Air pressure is low.
気圧が低い
氣壓低

Wing-tip vortex
翼端渦
翼尖漩渦

Air pressure is high.
気圧が高い
氣壓高

A vortex flow of air is made around the tip of the wing.
翼の先で渦状の空気の流れができる
在翼尖形成漩渦狀的渦流

V-formation
V字
V字型

A bird can fly easily by following the ascending air of the vortexes made by the bird flying ahead of it.
うしろに続く鳥は、前の鳥が作った渦が上昇するところにいると、楽に飛べる
後面的飛鳥，乘著前面的飛鳥製造的漩渦狀氣流而上升，輕鬆地飛行。

Q29 Fish don't close their eyes but don't they sleep?

① Some fish such as gray shark or goggle-eyed gobies
灰鯊   彈塗魚

have a membrane in their eyes, which takes the place of
薄膜   取代…

eyelids. ② Sunfish have muscles around the eye and they cover the
眼瞼   曼波魚   肌肉

eye, so they sometimes look as if they are blinking. ③ However,
看起來像是瞇著眼睛

most of fish do not have eyelids and they keep their eyes open.
持續地張開眼睛

④ You may think that fish do not sleep because they keep their eyes

open, but every fish does sleep.
實際上會睡眠

⑤ The time and place of sleeping depend on the kind of fish.
依照…   …的種類

⑥ Flounders, right-eyed flounders, and congers, which are active at
比目魚   鰈魚   星鰻   活動

night, sleep during the day, and parrot bass, filefish, and rockfish
石鯛   魨魚   石斑

sleep at the night.

⑦ The place of sleeping depends on the kind of fish, but they all

sleep in a place which is safe from attack such as space between
安全不被攻擊

rocks or in the sand.
在岩石縫隙或沙土中

⑧ Migratory fish like tuna are always swimming, so they sleep
迴游魚類   鮪魚

while swimming. ⑨ However, if they fell fast asleep, they would not
當牠們完全熟睡   是無法繼續游泳的

be able to swim. ⑩ They are swimming while sleeping lightly.
半夢半醒間

問29　魚は眼を閉じないけれど、寝ないの？

答 ①メジロザメ、ムツゴロウなど一部の魚は、**まぶたの役割をする膜**を持ちます。②マンボウは眼の周囲の**筋肉**が眼を覆い、**まばたきしているように見える**ことがあります。③しかし、ほとんどの魚にはまぶたがなく、**いつも眼を開けています**。④眠っているときも眼を開けているため、眠らないように見えますが、どんな魚も必ず眠ります。

⑤眠る時間や場所は、**種類によって異なります**。⑥夜行性の**ヒラメ、カレイ、アナゴ**などは昼間に眠り、**イシダイ、カワハギ、メバル**などは夜眠ります。

⑦眠る場所も魚の種類によって違い、**岩の間や砂の中**など敵に**攻撃されにくい場所**です。

⑧**マグロ**のように、常に泳ぎ続けている**回遊魚**は、泳ぎながら眠ります。⑨といっても、完全に眠ってしまうと**泳ぎ続けることなどできない**はずです。⑩半分眠ったうとうと状態で眠りながら泳いでいるのです。

問29　魚的眼睛無法閉闔，是不用睡覺嗎？

答 ①一些魚類諸如**灰鯊、彈塗魚**，他們的眼睛擁有類似眼瞼功能的**薄膜**。②曼波魚的眼睛周圍覆蓋著一圈**肌肉**，看起來像是瞇著眼睛看東西。③但是，幾乎所有的魚類並沒有眼瞼，**眼睛總是張開著**。④也因為睡眠的時候也張大眼睛，看起來像是沒有睡覺，實際上任何魚類都會睡眠。

⑤睡眠的時間、場所會依種類的不同而有所差異。⑥夜行性的**比目魚、鰈魚、星鰻**等等在日間睡眠，**石鯛、魨魚、石斑**等等則是夜間睡眠。

⑦睡眠場所也是依各種魚類而有所差異，通常是在**岩石縫、沙土中**等等敵人**不易攻擊**的場所。

⑧像**鮪魚**這種經常游泳的**迴游魚類**，則是一邊游泳一邊睡覺。⑨雖然如此，當牠完全熟睡的時候，是理所當然地無法繼續游泳的。⑩是在半夢半醒的狀態下，一邊游泳、一邊睡覺。

Q30 Why can salmon live both in the sea and the rivers?

A ①The important difference between fresh water and salt water for fish is osmotic pressure. ②Liquid flows from water with high osmotic pressure to that with low osmotic pressure. ③The osmotic pressure of seawater is higher than that of fresh water. ④The osmotic pressure of the body fluid of fresh water fish and seawater fish is almost the same; the pressure is just between that of seawater and fresh water. ⑤Therefore fresh water fish are in danger of swelling up because their body takes in water when they are in fresh water. ⑥One the other hand, seawater fish are in danger of shedding water too much when they are in salt water. ⑦It is the osmotic pressure adjustment function that controls this. ⑧Fresh water fish reduce the amount of water in their bodies by urinating a lot. ⑨On the other hand, seawater fish try not to urinate a lot and, at the same time, they give off salt through their gills.

⑩Fish which move from the sea to the rivers and from the rivers to the sea such as salmon and eel are called euryhaline fish. ⑪Euryhaline fish have an osmotic pressure adjustment function which allows them to live both in the sea and in rivers and can use it well in both types of water. ⑫Therefore they can live both in the sea and in rivers. (see p.90)

問30　サケはどうして、海と川の両方で生きていけるの？

答 ①魚にとって、**淡水と塩水の大きな違いは浸透圧**です。②**水分**は浸透圧が低いほうか
　　　　　淡水　海水　　　　　　　　　　　　　　浸透壓　　　　水分

ら高いほうへと移動します。③**海水と淡水を比べると、浸透圧が高いのは海水です。**④**淡水**
　　　　　　移動　　　　　　海水　　　　　　　　　　　　　　　　　　　　　　　　　淡水魚

魚と海水魚の体液の浸透圧はほぼ同じで、海水と淡水の中間の浸透圧です。⑤これが原因
海水魚　體液　　　　　　　　　　　　　　　　　　之間　　　　　　　　因素

で、淡水魚は淡水中にいるとき、体が水を吸収し水ぶくれ状態になる危険にさらされていま
　　　　　　　　　　　　　　　　　　吸収　　　　　　狀態　　　　危険

す。⑥逆に、海水魚は塩水中にいると、水分が流出してしまう危険があります。⑦これを調整
　　　　　　海水　　　　　　　　　　　　流失　　　　　　　　　　　　　　　　　調整

するのが、**浸透圧調整機能**です。⑧淡水魚は尿として多くの水分を排出し、体内の水分**を**
　　　　　調節機能　　　　　　　　　　　　　　　　排出　　　體內　水分

減らします。⑨海水魚は逆に水分をあまり出さないようにすると同時に、**エラから塩分を**
　　　　　　　　　　　　　　　　　　　　　　　　　　　同時　　　　　　　鹽分

排出しています。
排出

⑩**サケやウナギのように川から海へ、海から川へと回遊する魚は、広塩性魚**と呼ばれてい
　　　　　　　　　　　　　　　　　　　　　迴游　　　　　廣鹽性魚

ます。⑪広塩性魚は海水にも淡水にも対応する浸透圧調整機能を持ち、淡水と海水でこれ
　　　　　　　　　　　　　　適應

を使い分けることができます。⑫そのため、川でも海でも生きていけるのです。

問30　為什麼鮭魚可以同時生活在海洋與河川中？

答 ①對於魚類來說，**淡水與海水最大的差別就在於滲透壓**。②**水分**會由滲透壓低的一方朝高
的一方移動。③比較**海水**與淡水，滲透壓較高的為海水。④**淡水魚與海水魚的體液的滲透
壓幾乎相同，介於海水與淡水之間的滲透壓。**⑤因為這個因素，淡水魚生存在淡水中時，身體會
不停地吸收水分，會有**膨脹水腫**的危險。
⑥相反的，海水魚在海水當中生存時，會有水分不停地流失的危險。⑦能夠**進行調整**的，就是所
謂的**滲透壓調節機能**。⑧淡水魚會以尿液的方式排出多餘的水分，**減少體內的水分**。⑨相反的，
海水魚幾乎不排出水分、同時**由魚鰓排出鹽分**。
⑩像鮭魚或鰻魚這類由河川游入海洋、再由海洋游回河川的迴游魚類，稱之為**廣鹽性魚類**。⑪廣
鹽性魚類擁有可以同時適應海水與淡水的滲透壓調節機能，能夠分別在淡水與海水中使用。⑫因
此，不論是在河川或是海洋都能夠生存。

Q31 Don't lions become sick though they eat only meat, not vegetables?

①It is said that people who eat only <u>meat</u> are <u>likely to</u>
肉類　　容易

<u>suffer from</u> <u>lifestyle diseases</u>. ②So, you may <u>wonder if</u>
遭受（生病等等）…　生活習慣病　　　　　　　　　　　對…感到懷疑

<u>carnivorous animals</u> such as lions are <u>healthy</u>.
肉食動物　　　　　　　　　　　　健康

③It is known that animals in <u>the cat family</u> such as lions
貓科

<u>especially</u> eat only meat. ④Animals in the cat family have <u>a body</u>
特別是　　　　　　　　　　　　　　　　　　　　身體機制

<u>system</u> <u>which keeps them healthy</u> <u>through</u> eating only meat. ⑤One
保持健康　　　　　　透過…

of the reasons <u>lies in</u> <u>L-gulonolactone</u> <u>oxidase</u>. ⑥Since the meat-
在於…　L-古洛糖酸內酯　氧化酵素（氧化酶）

eating animals mainly in the cat family have this oxidase, they

can make <u>Vitamin C</u> from <u>grape sugar</u> in their bodies. ⑦<u>The</u>
維生素C　　　　葡萄糖

<u>nutrients</u> they can't make in their bodies such as <u>Vitamin A</u> <u>are</u>
營養素　　　　　　　　　　　　　　　　　　　維生素A

<u>taken in</u> by eating <u>grass-eating animals</u>. ⑧The meat-eating
藉由…攝取　　　　草食動物

animals can take in Vitamins <u>through</u> eating vegetables <u>digested</u>
透過…　　　　　　　適度消化的

<u>well</u> in <u>the stomach</u> and <u>the intestine</u> of the grass-eating animals
胃　　　　　腸

and through eating <u>the livers</u> of them.
肝臟

⑨The carnivorous animals don't need to eat <u>vegetables</u> and,
蔬菜

sometimes, they are <u>rather</u> <u>bad for</u> their health.
更正確地說　對…不好

問31 ライオンは肉ばかり食べて野菜を食べなくても、病気にならないの？

答 ①人間は、**肉ばかり食べていると生活習慣病にかかりやすくなる**といわれています。
人類　　　　　　　　　　　　　生活習慣病

②そこで気になるのが、ライオンなどのように肉ばかり食べている**肉食獣の健康**です。
肉食動物　健康

③ライオンなどの**猫科**の動物は、**特**に強い**肉食性**を持っていることが知られています。④猫
貓科　動物　　　特　　肉食性

科の動物は、肉ばかり食べていても**健康を保てる**ような**体のしくみ**を持っているのです。
健康

⑤その一例が**L-グロノラクトン酸化酵素**です。⑥猫科をはじめとする肉食獣はこの酵素を持
酸化酵素

っているため、**体内**で**ブドウ糖**から**ビタミンC**を合成することができます。⑦**ビタミンA**など
體內　　 合成

の体内で合成できないものは、**草食動物**を食べることによって補います。⑧たとえば、草
草食動物　　　　　　　　　獲得

食動物の胃や腸で**適度**に**消化**された植物や、草食動物の**肝臓**を食べることによりビタミン
適度　消化　　　　　　　　　　肝臓

を摂取します。
攝取

⑨肉食動物は、**植物を食べる必要**がないだけでなく、植物を食べるとかえって体調を悪く
植物　　需要　　　　　　　　　　　　　　身體

してしまう**可能性**もあります。
可能性

問31 獅子只吃肉類而不吃蔬菜，為什麼不會生病呢？

答 ①據說，人類若是只有用食用**肉類**，容易得到生活習慣病。②因此你會想知道像獅子等等這類、只有食用肉類的**肉食動物們的健康情形**。③獅子是我們已知的**貓科**等等動物當中，**特別擁有強烈肉食性的動物**。④貓科動物，擁有即使只有食用肉類也能**保持健康的身體機制**。⑤其中一個原因就是**L-古洛糖酸內酯氧化酶**。⑥由於肉食動物中以貓科動物為主，體內含有這種酵素、能夠將體內的葡萄糖合成為維生素C。⑦而體內無法合成的維生素A等等酵素，則藉由捕食**草食動物**而獲得。⑧舉例來說，可以透過食用在草食動物的**腸胃中適度消化**的植物、或是食用草食動物的**肝臟**來攝取維生素。⑨肉食動物不只不需要食用**植物**，甚至若是食用植物會有身體不適的可能性。

Why do sunflowers always face to the sun?

① It is only growing sunflowers that follows the direction of the sun. ② Growing sunflowers before their flowers come out face up into the sky at night, face east in the morning, and then, their head moves around to south and west following the movement of the sun. ③ This is the movement of the plants by a nature named phototropism.

④ Phototropism of plants such as sunflowers is the nature caused by a kind of growth hormones named auxin. ⑤ When this hormone reaches an amount, the plant speeds up its growth. ⑥ The reason why the plants face to the sun is that the auxin moves to the side of their stem which does not catch sunlight. ⑦ As a result, the amount of auxin is small on the side of the stem which catches sunlight and the amount of it on the side of the stem which does not catch sunlight is good enough to grow up. ⑧ The head of sunflower faces to the sun because only the side of the stem which does not catch sunlight speeds up its growth thanks to the difference in the amount of auxin between those sides.

問32　ヒマワリの花はどうして、太陽のほうをむいているの？

答 ①花が太陽の**方向**を追うように動くのは、成長期の**ヒマワリ**だけです。②花が**咲**
太陽　方向　　　　　　　　　　　　　　　　　生長期

く前の成長期のヒマワリは、夜は**上**をむき、朝になると**東**をむき、太陽の移動に合わ
移動

せて南から西の方向に頭を**回転**させます。③これは、植物の屈光性（くっこうせい）と呼ばれる動きの
轉動　　　　　　　　　　　　　　植物　趣光性

一つです。

④ヒマワリをはじめとする植物の屈光性は、**オーキシン**と呼ばれる**成長ホルモン**の一

種によるものです。⑤これが**適度な濃度**になると、植物は**成長を早めます**。
適當　濃度　　　　　　植物

⑥植物が太陽の方向をむくのは、オーキシンが茎の中で光が当たらない方向に移動す
移動

るからです。⑦**その結果**、茎の光が当たっている側のオーキシン濃度は低く、光があ
結果　　　　　　　　　　　　　　　　　　　　　　濃度

たらない側で成長に適した濃度になります。⑧ヒマワリの花が太陽の方向をむくの

は、オーキシンの濃度差により、光が当たらない側だけ成長が早まるからです。
濃度差異

問32　向日葵的花為什麼會朝著太陽的方向生長呢？

答 ①花生長的方向看起來是朝著太陽**的方向**追逐的，只有生長期的**向日葵**。②生長期的向日葵，在**開花**之前，夜晚**朝著天空**，到了白天則**面向東方**，隨著太陽移動的方向由南到西**轉動**它的頭部。③這可以稱之為植物**趨光性**的移動之一。④以向日葵為首的植物的趨光性，是來自於一種稱之為**苗長素**的成長荷爾蒙。⑤當這個**達到適當的濃度**時，植物**會加速生長**。⑥植物為什麼會朝向太陽的方向生長，是因為在**莖部當中**的苗長素移動到太陽照射不到的另一面所致。⑦**結果**，莖部照射到太陽的這一面濃度變低，照射不到太陽的這一面到達了適當的濃度。⑧向日葵的花朝向太陽方向的原因，是由於苗長素的濃度有了差異，只在太陽光照射不到的那一側生長加速所導致。

Salmons can control water and salt in them both in fresh water and in salt water.

サケは、海水中と淡水中で、体内の水分と塩分を調整する

鮭魚在海水與淡水之中，可以調整體內的水分與鹽分

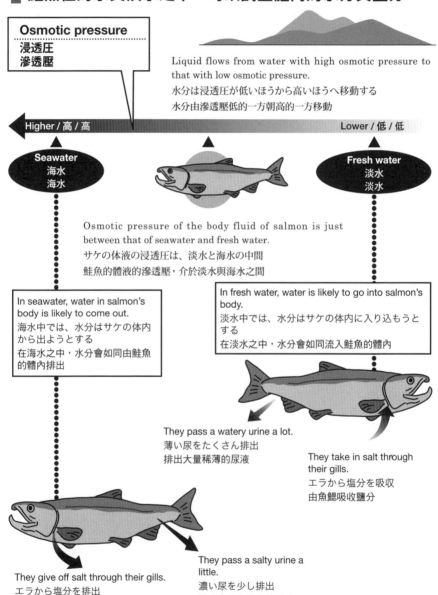

Osmotic pressure
浸透圧
滲透壓

Liquid flows from water with high osmotic pressure to that with low osmotic pressure.
水分は浸透圧が低いほうから高いほうへ移動する
水分由滲透壓低的一方朝高的一方移動

Higher / 高 / 高　　　　　　　　　　　　　　Lower / 低 / 低

Seawater
海水
海水

Fresh water
淡水
淡水

Osmotic pressure of the body fluid of salmon is just between that of seawater and fresh water.
サケの体液の浸透圧は、淡水と海水の中間
鮭魚的體液的滲透壓，介於淡水與海水之間

In seawater, water in salmon's body is likely to come out.
海水中では、水分はサケの体内から出ようとする
在海水之中，水分會如同由鮭魚的體內排出

In fresh water, water is likely to go into salmon's body.
淡水中では、水分はサケの体内に入り込もうとする
在淡水之中，水分會如同流入鮭魚的體內

They pass a watery urine a lot.
薄い尿をたくさん排出
排出大量稀薄的尿液

They take in salt through their gills.
エラから塩分を吸収
由魚鰓吸收鹽分

They give off salt through their gills.
エラから塩分を排出
由魚鰓排出鹽分

They pass a salty urine a little.
濃い尿を少し排出
排出少量有鹽分的尿液

The side of the sunflowers' stem which does not catch sunlight grows well.

ひまわりの茎（くき）は、日陰側が活発に成長する
向日葵的莖部，在日陰處的那一側生長地較好。

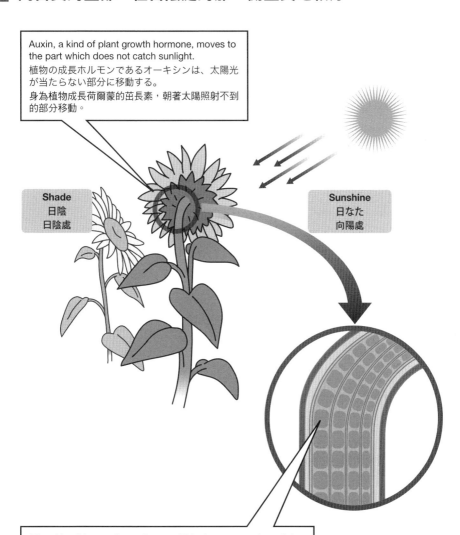

Auxin, a kind of plant growth hormone, moves to the part which does not catch sunlight.
植物の成長ホルモンであるオーキシンは、太陽光が当たらない部分に移動する。
身為植物成長荷爾蒙的苗長素，朝著太陽照射不到的部分移動。

Shade
日陰
日陰處

Sunshine
日なた
向陽處

The side of the sunflowers' stem which does not catch sunlight and has much auxin grows fast because it undergoes cell division actively.
オーキシンが多い日陰側は細胞分裂が活発なので、成長が早い。
由於苗長素較多的日陰側，細胞分裂較為活潑，成長速度較快。

Q 33 Why don't penguins live in the North Pole?

①You will probably be surprised to hear that penguins
吃…一驚   企鵝
used to live at the North Pole. ②In fact, great auks living
過去曾是…   北極   大海雀
in the North Pole were commonly called "penguin" in the old days.
                         俗稱為…

③Unfortunately, people hunted them too much and they died out
不幸地是…   獵捕…   絕種
in the 1840's.
1840年代

④Around that time, Europeans often went to look around the
                        歐洲人   探險
southern hemisphere and they found a new kind of bird living
南半球
around cold seaside in the southern hemisphere. ⑤Since the birds
住在寒冷的海邊
looked like the great auks, they came to be called penguin. ⑥They
看似…   以…的名稱來命名
are the penguins we know today.

⑦Academically, these two kinds of birds are not closely related
學術上的   並不屬於近親
because great auks are Charadriiformes and penguins are
                        鷸形目的
Sphenisciformes. ⑧However, great auks were like penguins today.
企鵝目的
⑨Both great auks and penguins could not cross the equator
                                                跨越… 赤道
because they did not like heat. ⑩Scientists think that these
                        熱
academically different birds became similar in the course of
學術上相異的…   相似   在進化的過程中
evolution in a similar environment.
              在相似的環境下

92

問33　ペンギンはどうして、北極にはいないの？

答 ①「**ペンギンは、もともと北極にいた鳥**」だと聞くと、みなさんは**驚く**に違いありませ
北極

ん。②実は「ペンギン」はもともと、北極にいた**オオウミガラス**につけられた**俗称**でした。③
俗稱

残念なことに、この鳥は人間に乱獲され、1840年代に絶滅してしまいました。
可惜　　　　　　　濫捕　　　　　　　絕種

④ちょうどその頃、**ヨーロッパ人**による**南半球の探検**が盛んになり、南半球の寒い**海辺に生**
南半球　探險　　　　　　　　　　　　　　海邊　棲息

息する新種の鳥が発見されました。⑤この鳥はオオウミガラス**に似ていた**ためペンギンと呼
新種　　發現

ばれるようになりました。⑥これが**現在**のペンギンです。
現在

⑦オオウミガラスは**チドリ目**であるのに対して、ペンギンは**ペンギン目**であり、この2種類の
もく　　　　　　　　　　　　　　　　　　　　　　　　　　種類

鳥は分類上は**近い仲間とはいえません。**⑧にもかかわらず、オオウミガラスは現在のペンギ
分類　　　近親

ンと同じような姿をした鳥でした。
外型

⑨ペンギンもオオウミガラスも、**暑さが苦手であるため赤道を横切って移動できなかったの**
不耐　　　　　跨越赤道

でしょう。⑩その一方で、分類上まったく異なる鳥が似た**環境のもとで進化**するうちに、**似**
分類　　　　　　　　　　　　　環境　　　進化

た姿になったものと考えられています。

問33　為什麼北極沒有企鵝？

答 ①要是聽到「企鵝原本是生存在北極的鳥類。」，各位勢必會大吃一驚。②事實上「企
鵝」原本是生長在北極的鳥類俗稱為大海雀。③**可惜的是**，這種鳥類因為人類的濫捕，已
經在1840年代絕種了。④約在那個時期，歐洲人興起了到**南半球的探險**，發現了棲息在南半球寒
冷**海邊**的新種鳥類。⑤由於這種鳥類與大海雀**極為相似**，因此便以大海雀的俗名─企鵝來命名。
⑥這就是現在的企鵝。⑦相對於大海雀為**鴴形目鳥類**，企鵝則是**企鵝目鳥類**，這2種鳥類在分類上
並**不屬於近親**。⑧雖然如此，大海雀卻是擁有與今日的企鵝相似外型的鳥類。⑨不論是大海雀還
是企鵝，都因為牠們不耐**酷熱**而無法跨越赤道。⑩另一方面，科學家認為，在分類上完全不同的
鳥類會成長得如此相似，是由於**在相似的環境下進化**所導致。

Q34 Why do ants walk in line?

①You sometimes see ants walking in line in parks or in your garden. ②The ants' line is sometimes more than 50 meters long. ③An ant follows the ant ahead of it even when they walk a long way.

④Ants walk in line mainly when they go out for food. ⑤An ant which has found food comes back to its nest with the food and the ant lets go a material from its body as a landmark on his way back home. ⑥This landmark is a material called pheromone. ⑦When the ant which has found food comes back to its nest, other ants in the nest start to go to the place in which there is food following the landmark pheromone. ⑧This forms the ants' line. ⑨Ants can do almost nothing with the help of their vision because their vision is very poor. ⑩Instead, they use their antennae to touch something and find out about it, and also use pheromone as a means of sending messages. (see p.100)

問34 アリはどうして、行列をつくって歩くの？

答 ①公園や庭で、**アリが行列をつくって歩いている**ところを見かけることがあります。②長さは時には**50メートルを超える**こともあります。③そんなに長い距離でも、後ろを歩くアリは前を歩くアリが歩いたあとをたどります。

④アリが行列をつくるのは、**主に餌を探しに行くとき**です。⑤1匹のアリが餌を見つけると、腹部から**目印になる物質を出し**ながら餌を抱えて**巣**に戻ります。⑥目印となるのは、**フェロモン**と呼ばれる物質です。⑦餌を見つけたアリが巣にたどり着くと、巣にいたアリたちは帰ってきたアリのフェロモンをたどり、餌の場所を目指します。

⑧これがアリの行列です。

⑨アリは目がそれほど発達していないので、**視覚に頼ることはほとんどできません。**
⑩それを補うのが、ものに触れて確認するための**触角**と、**情報伝達の手段としての**フェロモンなのです。

問34 螞蟻為什麼會排成一列行走？

答 ①在公園或庭院裡，經常可以發現**螞蟻排成一列**的行走。②排列的長度經常都**超過50公尺**。③即使這麼長的距離，後面行走的螞蟻還是緊緊跟著前面行走的螞蟻的腳步。④螞蟻會呈行列行走，**主要是外出尋找食物**。⑤當1隻螞蟻發現了食物，會一面由腹部釋**放出做為信號的物質**、一面將食物抱回**巢穴**。⑥這個信號是一種稱為**費洛蒙**的物質。⑦當發現了食物的螞蟻一回到巢穴，在巢穴裡的螞蟻便依照返巢的螞蟻的費洛蒙，去尋找食物所在的場所。⑧這樣就形成了螞蟻的行列。⑨由於螞蟻的眼睛不怎麼發達，幾乎**無法依靠視覺活動**。⑩取而代之的是，藉由接觸以達到確認的**觸角**、以及做為**情報傳達手段**的費洛蒙。

Q35 Why aren't deep-sea fish crushed under water pressure?

① Fish living at a depth over 200 meters are generally
<u>住在水深200公尺以下</u>
called <u>deep-sea fish</u>. ② <u>Water pressure</u> at a depth of 200
深海魚　　　　　　水壓
meters is about <u>20 bar</u>. ③ An <u>empty</u> <u>PET bottle</u> (Polyethylene
20巴　　　　　　空的　保特瓶　　　　聚乙烯對苯二甲酸酯瓶
terephthalate bottle) <u>with a cap</u> <u>sunk</u> to this depth <u>will be crushed</u>
蓋上瓶蓋　潛入…　　　　　　在…之下會被壓扁
<u>flat under</u> this pressure. ④ <u>Then</u>, why aren't deep-sea fish crushed
然而
under such pressure.

⑤ Fish living in <u>shallow</u> water such as <u>horse mackerel</u> and <u>sardine</u>
淺的　　　　　　　　　竹筴魚　　　沙丁魚
have <u>air bladders</u> which <u>control</u> their <u>ups and downs</u> <u>according to</u>
氣囊　　　　　　調整…　　　上昇與下潛　根據…
<u>the amount of</u> air in them. ⑥ The <u>specific gravity</u> of sea fish is
…的量　　　　　　　比重
higher than that of <u>the seawater</u>, so they can not <u>come up</u>
海水　　　　　　　　　　　　上昇
<u>without</u> the air bladders. ⑦ Then, <u>what would happen</u> if horse
在沒有（或缺少）的情況下　　　　會發生什麼事呢？
mackerel or sardine <u>went down</u> to the deep sea? ⑧ <u>They would die</u>
假如潛入　　　　　　　　　　可能死亡
because their air bladders would be crushed. ⑨ One of the reasons
that deep-sea fish can live <u>under high pressure</u> <u>lies in</u> their air
在強大的水壓下　　　憑藉著…
bladder. ⑩ The air bladder of deep-sea fish is <u>filled with</u> fat, and, <u>in</u>
充滿了…　　　　脂肪
<u>other points</u>, they are also <u>designed to</u> live under high pressure.
另一個重點是　　　　　　為了…而設計成
⑪ Some deep-sea fish do not have an air bladder. ⑫ However, the
secret of deep-sea fish is not well <u>understood</u> <u>yet</u>. ⑬ Their
尚未…　　理解
<u>mysteries</u> will be <u>solved</u> <u>in the near future</u>.
神秘　　　　　　被解開　　不久的將來

問35　深海魚はなぜ、水圧に押しつぶされないの？

答 ①**深海魚**とは、一般的に**水深200メートル以下**の深海に**生息する魚**を指します。②水深200メートルの**水圧**は約**20気圧**。③**空のペットボトル**にふたをして沈めると、ペシャンコになってしまうほどの力を受けます。④深海魚はこれだけの**圧力**を受けているにもかかわらず、なぜつぶれないのでしょうか？

⑤**アジやイワシ**をはじめとした**浅い海**に暮らす魚は、**空気の量**で**浮き沈み**を**調整する**うきぶくろを持っています。⑥**海水魚**の**比重**は**海水**より重いため、うきぶくろがないと**浮上する**ことができません。

⑦では、**もしもアジやイワシが深海に行ったら**どうなるでしょうか？⑧うきぶくろがつぶれて**死んでしまう**でしょう。⑨深海魚が**強い水圧に耐えられる秘密**の一つは、うきぶくろにあります。⑩深海魚のうきぶくろは、空気の代わりに**脂肪で満たされる**など、高い水圧にも耐えられるような**構造**をしています。⑪なかには、うきぶくろを持たないものもいます。⑫ただし、深海魚についての**研究**はそれほど進んでいません。⑬**今後**、多くの**謎**が解き明かされることでしょう。

問35　為什麼深海魚類不會因為強大的水壓而被壓扁？

答 ①說到**深海魚**，一般是指棲息在**水深達200公尺以下**深海的魚類。②水深200公尺的**水壓**約為**20巴**。③將一個蓋上瓶蓋的**保特瓶**潛入這個深度，在這樣的壓力下**會被壓扁**。④那麼，為什麼深海魚雖然承受了這樣大的壓力，卻不會被壓扁呢？⑤以**竹筴魚或沙丁魚**這類生存在淺海的魚類來說，牠們擁有可以**調節空氣量、調整上升與下潛的氣囊**。⑥由於海水魚的**比重比海水重**，沒有氣囊便沒有辦法上升。⑦那麼，要是**竹筴魚或沙丁魚潛入深海裡**會如何呢？⑧牠們會因為氣囊被擠破而**死亡**。⑨深海魚能夠**承受強大水壓**的祕密之一，就在於氣囊。⑩深海魚的氣囊裡，取而代之空氣的是**充滿了脂肪**，並且擁有能夠抵抗強力水壓的構造。

⑪也有些深海魚，並沒有氣囊的構造。⑫但是，有關深海魚的研究目前還不夠深入。⑬**不久的將來**，將會有許多謎團陸陸續續地被揭開。

Q36   Why do leaves turn red in autumn?

① As the temperature falls, plants don't work actively and
氣溫　　　　　　　　　下降　植物

the amount of water taken in from their roots becomes
水分的量　　　　　　　由…吸收　　　　　　根部

less. ② The leaves die from lack of water when the water taken in
減少　　　　　　因為…而枯死　不足

from the roots evaporates from the leaves. ③ Therefore trees shed
蒸發　　　　　樹葉　　　　　　　　　　　　脫落…

their leaves fall and get ready for the cold season.
為…作準備

④ A separation layer is formed between the leaves and the branch
剝離層　　　　　　被形成…　　　　　　　　　　　　　　樹枝

when the temperature starts falling. ⑤ The leaves look green
看起來

because of the chlorophyll in them needed for photosynthesis
葉綠素　　　　　　　需要…　　　光合作用

⑥ This chlorophyll is broken down because the separation layer
被分解

blocks the water and the nutrients.
被…阻礙　　　　　養分

⑦ In the case of maple, the plants whose leaves turn red, the
楓葉　　　　　　　　　　　　　　轉為紅色

separation layer keeps sugar made at the leaves from going into
避開…　糖分

the branch. ⑧ Therefore anthocyanin, a coloring matter of red, is
花青素　　　色素

made up from the sugar and the leaves turn red.
由…合成

⑨ The leaves of ginkgo and so on turn yellow because of carotenoid
銀杏　　　　　　　　　　　　　　　　類胡蘿蔔素

a yellow coloring matter. ⑩ The color of carotenoid, which is always

in leaves, comes to stand out because the chlorophyll is broken
引人注目

down.

問36 秋になると、なぜ葉っぱが赤くなるの？

答 ①温度が下がると植物の活動が鈍り、根が吸い上げる水の量が減ります。②根が吸い
植物　活動　遅緩　　　　　　　　吸

上げた水分が葉から蒸発してしまうと、水分が不足して枯れてしまいます。
蒸發　　　　　　　　　　　　不足

③そこで木は、葉を落として、寒い季節に備えます。

④気温が下がり始めると、葉と枝の間に離層が形成されます。⑤木の葉が緑色に見えるの
氣溫　　　　　　　　　　　　　剥離層　形成　　　　　　　　　　　　緑色

は、植物が光合成を行なうのに必要な葉緑素があるからです。⑥しかしながら、離層がで
植物　光合作用　　　必需要　葉緑素　　　　　　　　　　　　　　剥離層

きると、それが水分や養分の流れを妨げるため、この葉緑素が分解されていきます。
養分　　　　　　　　　　　　　　　　分解

⑦モミジのように赤く色づく植物の場合、葉でつくられた糖分が離層によって枝に送られな
類型　　　　　　　　　　　　　　　糖分

くなります。⑧その結果、この糖分からアントシアニンという赤い色素が合成されて、木の葉
色素　合成

は赤くなります。　⑨イチョウなどで葉が黄色くなるのは、カロチノイドという黄色の色素に
黄色　色素

よるものです。⑩カロチノイドはもともと葉に含まれていますが、葉緑素が分解されることに
分解

より、カロチノイドの黄色が目立ってくるわけです。
醒目

問36 到了秋天，樹葉為什麼會變成紅色？

答 ①當氣溫下降的時候，植物的活動會變遲緩、從根部吸收的水量會減少。②當由根部吸收的水分經由樹葉蒸發掉，水分不足就會枯死。③因此，樹木就會開始落葉，準備度過寒冷的季節。④氣溫開始下降的時候，在樹葉與樹枝之間，會形成剝離層。⑤樹葉看起來是綠色的，是因為裡面含有植物進行光合作用所必需要的葉綠素。⑥然而，剝離層形成後會阻礙水分與養分的流動，葉綠素就會分解。⑦而像楓葉這類轉為紅色的植物，是由於樹葉製造的糖分因為剝離層的隔離而無法傳送到枝幹。⑧結果，由糖分合成稱為花青素的紅色色素，讓樹葉變成紅色。⑨而像銀杏這類樹葉轉為黃色的植物，則是由於稱為類胡蘿蔔素的黃色色素。⑩類胡蘿蔔素原本就存在於樹葉之中，因為葉綠素被分解，使得類胡蘿蔔素的黃色趨於醒目。

Ants use a pheromone another ants let go as a landmark.

アリはほかのアリが出したフェロモンを目印にする
螞蟻對其它螞蟻放出費洛蒙做為信號

Finding food
餌をみつける
發現食物

An ant which has found food comes back to its nest releasing a pheromone from its body.
餌を見つけたアリは、腹部からフェロモンを出しながら巣に戻る
發現食物的螞蟻，一面由腹部釋放出費洛蒙、一面返回巢穴。

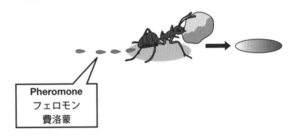

Pheromone
フェロモン
費洛蒙

Ants in the nest start to go to the place in which there is food following the landmark pheromone.
巣にいたアリがフェロモンをたどり、餌集めに出発する
在巢穴的螞蟻，朝著信號費洛蒙去尋找食物。

A lot of ants go to get food following the pheromone and form the line.
多くのアリがフェロモンをたどって餌集めに行くため、行列になる
大量的螞蟻隨著費洛蒙去覓食，形成了行列。

The leaves turn red because a red coloring matter is made.

木の葉が紅葉するのは、赤い色素ができるから
樹葉會變成紅色，是因為紅色色素的原故

How the leaves turn red
木の葉が紅葉するしくみ
樹葉如何變成紅葉

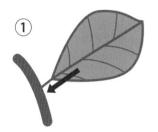

Starch is made by chlorophyll and sent to the trunk.
葉緑素によりでんぷんが作られ、幹に送られている
由葉綠素製作出澱粉，運送到枝幹

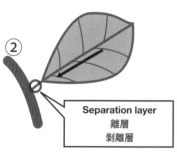

Separation layer
離層
剝離層

Chlorophyll doesn't work well and a separation layer is formed at the root of a leaf when it becomes cold.
寒くなると、葉緑素の働きが弱くなり、葉のつけ根に離層が形成される
天氣變涼，葉綠素的活動緩慢，在樹葉的根部形成了剝離層

Nutrients stay in the leaves because the separation layer keeps them from entering the trunk. These nutrients are broken down into the sugar.
離層により幹に流れなくなった養分が葉にたまる。それが分解されて糖分になる
由於剝離層阻斷，養分無法進入枝幹而囤積在樹葉。這些養分分解之後形成糖分

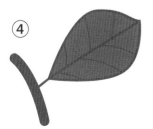

Anthocyanin, a red coloring matter, is made up from the sugar and the leaves turn red.
糖分から赤い色素であるアントシアニンが合成され、木の葉が赤く色づく
由糖分合成為紅色色素的花青素，使得樹葉變成紅色

There are many different creatures on the earth.

地球は生き物の宝庫です
地球是生物的寶庫

Lion
ライオン
獅子

Male and Female
男女
男女

Elephant
象
大象

Giraffe
キリン
長頸鹿

Cat
猫
貓

Tiger
トラ
老虎

Mouse
ねずみ
老鼠

Chicken
ニワトリ
雞

Panda
パンダ
熊貓

Seal
アザラシ
海豹

Sunflower
ひまわり
向日葵

Tulip
チューリップ
鬱金香

Questions about human body

第４章

人間の体の疑問
人體的疑問

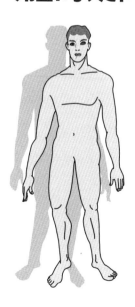

Q37 Did human beings evolve from apes?

① Although all apes are grouped together, there is a bigger difference between chimpanzees and Japanese monkeys than between chimpanzees and human beings. ② It is chimpanzees that human beings are close to on the evolutionary tree. ③ It is said that, about 5 million years ago, chimpanzees and human beings separated from a kind of ape which has died out, and each of them evolved differently.

④ For a long time scientists could not find out what kind of ape the common ancestor of human beings and chimpanzees was. ⑤ However, scientists today think that one of the anthropoids whose fossils have been found such as Nakalipithecus and Ouranopithecus, is the common ancestor of human beings and chimpanzees.

⑥ In the study of genes, the difference between chimpanzees and human beings in DNA (deoxypentose nucleic acid sequence) is within several percentage points. ⑦ This difference is very small in genetics. ⑧ This small difference in genes is a result of the different evolutions for as long as 5 million years.

⑨ Therefore it is wrong to say that human beings evolved from apes. ⑩ You should say that apes and human beings have a common ancestor. (see p.116)

問37 ヒトはサルから進化したの？

答 ①ひと言で**サル**といっても、**チンパンジーとニホンザル**では、チンパンジーとヒト以上の違いがあります。②**進化の系統**から見てヒトと近いのは、チンパンジーです。③チンパンジーとヒトは、現在はいないサルと**500万年**ほど前に枝分かれし、ヒトとチンパンジーに分かれて**進化した**ということです。

④ヒトとチンパンジーの**共通の祖先**がどのようなサルであったのか、**長い間**わかりませんでした。⑤最近では、**ナカリピテクス、オウラノピテクス**など、**化石**で発見された**類人猿**が、ヒトとチンパンジー共通の祖先と考えられています。

⑥**遺伝子**の研究では、チンパンジーとヒトとの**DNA配列**の違いは、**数パーセント以内**とされています。⑦この違いは、**遺伝学**的に見るとほんのわずかにすぎません。⑧このわずかな遺伝子の違いが500万年間にわたり、違う**進化**をしてきた**結果**なのです。

⑨したがって、「ヒトはサルから進化」したというのは正しくありません。⑩「サルとヒトの祖先は同じだ」というべきなのです。

問37 人類是由猴子演化而來的嗎？

答 ①簡單的說，雖然都是**猴子**，但是比較**黑猩猩**與**日本獼猴**、以及黑猩猩與人類就有很大的不同。②從**演化樹**來看，人類比較接近的是黑猩猩。③據說，黑猩猩與人類是由現在已經滅絕的猴子，在**500萬年**前左右**分支**、各自演化成不同物種。④人類與黑猩猩的共同祖先究竟是哪一種猴子，經過**相當長一段時間**還是無法發現。⑤直到最近，經由化石所發現的**仲山納卡里猿、歐蘭猿**等等類人猿，被認為是人類與黑猩猩共同的祖先。⑥經**遺傳因子**的研究，黑猩猩與人類的DNA序列差異，僅在**數個百分比之內**。⑦這樣的差異，在**遺傳學**上來說，幾乎是微乎其微。⑧這樣細微的遺傳因子差異，是經過了500萬年的**演化**差異後所形成的**結果**。⑨因此，「人類是由猴子演化而來」的說法並不正確。⑩應該說「猴子與人類擁有共同的祖先」才對。

Q38　Why must all human beings die?

[1] Some scientists say that telomere has something to do with the life span of human beings. [2] Telomere is the tail-like part at the ends of chromosomes. [3] Scientists say that telomere protects DNA, and that it is also necessary for cells to divide in order to make new normal cells.

[4] Most of the cells of living things divide to make new cells again and again. [5] Telomeres shorten every time cells divide. [6] When telomeres become short after cells divide a certain number of times, the cells stop dividing. [7] It is thought the reason for this is that cells keep themselves from dividing to make abnormal cell, such as cancer cells, due to the fact that telomeres do not work well. [8] The state at which cells stop dividing is called cell aging.

[9] Today, it is said that the life span of human beings is about 120 years at longest. [10] There are quite a few factors which decide the life dpan of human beings, and scientists think telomere is one of them.

問38　人間はなぜ、必ず死ぬの？

答　①人間の寿命にテロメアがかかわっている、という説があります。②テロメアと

いうのは、**染色体の両端にあるしっぽのような部分**です。③このテロメアはＤＮＡを

守るだけでなく、**正常に細胞分裂を行なうなどのために必要なもの**とされています。

④**生物の多くの細胞**は、分裂を繰り返しています。⑤テロメアは細胞**分裂のたびに短く**

なっていきます。⑥そして、**一定回数**の細胞分裂を行ない、ある長さより短くなる

と、細胞は分裂をやめてしまいます。⑦これは、テロメアが十分に働かないことによ

り、癌化など、**異常な**細胞分裂を**起こすことを防ぐ**ためと考えられています。⑧細胞

が分裂をやめた**その状態**が「**細胞の老化**」です。

⑨現在、人間の寿命は最長120歳前後とされています。⑩もちろん、人間の寿命を左右

する**要因**は単純なものではありませんが、テロメアもその一つではないかと考えられ

ています。

問38　人類為什麼一定會死亡？

答　①據說，人類的壽命與**端粒**有著相當關連。②所謂的端粒，是指位在**染色體的兩端**、狀似**尾巴的部分**。③端粒不只能夠**保護DNA**，也是細胞進行**正常的細胞分裂**所必要的結構。④**生物大部分的細胞**，會不斷地進行細胞分裂。⑤端粒在每次的細胞**分裂後，都會縮短**。⑥當細胞進行一定次數的細胞分裂後，端粒會**縮短**到某個程度、細胞則停止分裂。⑦據說，因為端粒停止了動作，就**能防止**引起癌細胞等等這類**異常**的細胞分裂發生。⑧阻止細胞分裂的**狀態**，稱為「**細胞老化**」。⑨今日，據說人類最長的壽命在120歲左右。⑩當然，左右人類壽命的**要素**不會只有單單一項，端粒也只是其中之一而已。

Q39 Why isn't the human body covered with hair like dogs or cats?

①There are some unproved theories about why the
假說
human body is not covered with hair like most of the
人類的身體　　　　　沒有覆蓋著毛髮
other mammals.
哺乳類

②One of them is the theory that human beings lost their hair to
學說　　　　　　人類
control their body temperature. ③When the ancestors of human
調整…　　　　體溫　　　　　　　　祖先
beings lived in trees, they were not exposed to the direct sunlight.
暴露　　　　　　直射陽光

④Later, when they began to walk upright and live on grasslands,
之後　　　　　　　　開始　　　直立步行　　　　　　　　草原
they had to work under the burning sun. ⑤The theory is that, in
在炎熱的太陽下
order to adapt themselves to these new surroundings, the sweat
為了去適應…　　　　　　　　　　　　　　環境　　　　汗腺
glands of human beings developed to control their body temperature
發達
by sweating and that they lost their hair at the same time.
藉由流汗　　　　　　　　　　　　　　　同時

⑥Another one is called "Human Neoteny Theory." ⑦Neoteny
幼態演進說
means that animals grow old with the characteristics they have at
保留特質　　　　　　　他們出生的時候
birth. ⑧Newborn chimpanzees have no hair. ⑨This theory says
剛誕生的黑猩猩
that human beings grow old without hair like the newborn
沒有…
chimpanzees.

⑩There are other theories about this matter but none of them have
這個疑問　　　　完全沒有…
been confirmed yet.
還沒有被確認

問39　ヒトはどうして、イヌやネコのように全身が毛で覆われていないの？

答　①ヒトがほかの**哺乳類**のように全身を体毛で覆われていない理由については、いくつかの**説**があります。

②まず、**体温を調整する**ために、体毛を失ったとする**説**です。③ヒトの**祖先**が木の上で生活していたころには、**直射日光**にさらされることはありませんでした。④**直立歩行**を始め、草原で暮らすようになると、**炎天下**で活動しなければならなくなります。⑤この**環境**に合わせて、汗で体温調整ができるように**汗腺**を**発達**させ、**同時**に体毛を失ったとするものです。

⑥もう一つは、「**人類ネオテニー説**」と呼ばれるものです。⑦ネオテニーとは生まれたばかりの**性質**を残したまま**成熟**することです。⑧**生まれたばかりのチンパンジーには体毛があります**。⑨ヒトはその状態で大人になった**生物**だとするのです。⑩ほかにも**諸説**ありますが、まだ**定説**とされるものはありません。

問39　人類為什麼不會像貓、狗一樣，全身覆蓋著毛髮？

答　①關於人類不如其他哺乳類一樣全身覆蓋著毛髮的理由，有以下幾個不同的假說。②首先是，為了調整體溫而失去體毛的學說。③當人類的祖先生活在樹上的時候，並沒有暴露在直射陽光下。④當開始直立步行的時候，變成生活在草原上，就不得不在炎熱的太陽底下活動。⑤配合這樣的環境，能夠藉由流汗調整體溫的汗腺變得發達、同時也失去了體毛。⑥還有一個理論被稱之為「幼態演進說」。⑦幼態的意思是，保留剛出生的特質而成長。⑧甫誕生的黑猩猩也沒有體毛。⑨這個學說認為，人類就是維持初生的狀態而長大成人的物種。⑩關於這個疑問還有其他諸說，但是還未有有力的論點。

Q40 Why do you shed tears when you are sad?

①The surface of the eye is always covered with thin layer
表面　　　　　　　　　　覆蓋著…　　　　薄薄一層
of tears. ②This is because tears reduce the effect of
眼淚　　　　　　　　　　　　　　減少
stimuli from outside and protect the eye. ③You shed tears when
刺激　　　　　　　　　　　　　　　　　　　流淚
you have something in your eye and it's surface is stimulated.
　　　　　　　　　　　　　　　　　　　　　　遭遇刺激

④This is because the trifacial nerve is working.
　　　　　　　　三叉神經

⑤Tears which come in the eye when you are sad or happy are
　　　　　　流出（眼淚）
different from tears which come in the eye when it is stimulated.

⑥This is because the autonomous nerve is working. ⑦The
　　　　　　　　自律神經
autonomous nerve is made up of the sympathetic nerve and the
　　　　　　　　由…構成　　　交感神經
parasympathetic nerve. ⑧When you are very much excited by
副交感神經　　　　　　　　　　　　　　　　因…而激動
anger or regret, the sympathetic nerve works actively and a small
憤怒　　懊悔　　　　　　　　　　　　　　　　　　　少量的
amount of salty tears comes in the eye. ⑨When you are sad or
　　　　鹹的
moved, the parasympathetic nerve works actively and a large
感動　　　　　　　　　　　　　　　　　　　　　大量的
amount of watery tears comes in the eye.
　　　　充滿水的

⑩In tears, there are prolactin and adrenocorticotropic hormones,
　　　　　　　　　　催乳激素　　　　促腎上腺皮質激素
which cause a state of tension under stress, and manganese, a
　　　引起…　緊張狀態　　在壓力之下　　錳
large amount of which is said to cause depression. ⑪Some
　　　　　　　　　　據說…　　　憂鬱症
scientists say that you feel good after you shed tears because these
substances are passed out of the body with tears.
　　　　由…排出

問40　なぜ、悲しいと涙が出るの？

答 ①眼の**表面**にはいつも、**少量**の**涙で覆われています**。②眼が受ける**刺激をやわらげ**、眼を守るためです。③眼にごみが入ったときや、眼の表面が**刺激を受けたとき**にも**涙が出ま**す。④これらは**三叉神経**の働きによるものです。

⑤悲しいときやうれしいときなどに出る涙は、刺激を受けたときに出る涙とは違います。

⑥それは**自律神経**によるものです。⑦自律神経は、**交感神経**と**副交感神経**に分かれています。⑧**怒りや悔しさ**などの強い感情の**昂ぶり**を感じたときには、交感神経のほうがよく働き、**しょっぱい涙が少し**出ます。⑨悲しいときや**感動**したときは、副交感神経のほうがよく働き、薄くて**水っぽい涙が大量に**出ます。

⑩涙には**プロラクチン**や**副腎皮質刺激ホルモン**など、**ストレスに反応して緊張**を誘発する物質や、**過剰**になるとうつ病の原因になるとされている**マンガン**が含まれています。⑪泣くとすっきりするのは、これらが涙とともに**排出される**からだという説もあります。

問40　為什麼悲傷的時候會流眼淚？

答 ①眼球的**表面**總是覆蓋著少量的**眼淚**。②目的是緩和眼球遭受到**外力的刺激**，以及保護眼球。③當眼球有異物進入時，或是眼球表面**遭遇到刺激時，就會流淚**。④這是由於**三叉神經**運作的緣故。

⑤在悲傷的時候或高興的時候等等流淚，則是跟遭遇到刺激時所流淚的情況不同。

⑥那是由於**自律神經**運作的緣故。⑦自律神經是由**交感神經**與**副交感神經**所構成。⑧當**憤怒**或是**懊悔**等等感受到強烈的情緒發生時，交感神經這邊會開始活躍、流出**少量的鹹**的眼淚。⑨而悲傷或**感動**的時候，副交感神經這邊則會開始活躍、流出**大量含有水分**的眼淚。

⑩在眼淚之中含有因為**反映壓力、緊張狀態**下產生的**催乳激素、促腎上腺皮質激素**等物質，以及堆積過剩的話會形成**憂鬱症**原因的**錳**。⑪科學家認為，當你哭泣之後心情會舒暢，是由於這些物質隨著眼淚**排出體外**的原故。

Q41　Why can you drink many glasses of beer in a short time?

①You may be able to drink a few big glasses of beer
大啤酒杯
without difficulty but it may not be easy for you to drink
不困難地　　　　　　　　　　　　　（人）作…是…
the same amount of water in the same length of time. ②Why?
等量的　　　　　　　　　相同時間
③When you drink water, it stays in the stomach for a while and it
胃
flows slowly to the duodenum, the small intestine, and the large
流動　慢慢地　　十二指腸　　　　小腸　　　　　　　　　大腸
intestine, and then it is taken into at the large intestine. ④You can
吸收
not drink a large amount of water in a short time because the
大量的
amount of water which flows from the stomach to the duodenum
can not keep up with the amount of water you drink, so the stomach
無法跟得上
becomes full of water.
變成充滿
⑤Beer has ethyl alcohol in it. ⑥Alcohol is likely to be taken in
酒精　　　　　　　　　　　容易…
through the mucous membranes. ⑦It is thought that, because of
黏膜
the nature of alcohol, 20 to 30 percent of the amount of beer is
性質
taken into the stomach and the duodenum, in which water is not
taken in at all. ⑧Beer is not only taken in easily but also it has
不僅…　　　　　　　　而且…
two kinds of diuretic effects. ⑨One of these effects is that alcohol
利尿作用
cuts down on the production of a hormone which controls the
減少…　　　分泌　　　　　荷爾蒙　　　抑制…
amount of urine. ⑩The other is the diuretic effect the potassium in
尿　　　　　　　　　　　　　　　鉀
beer has. ⑪You can drink a large amount of beer in a short time
because beer is taken in easily and it is excreted from the body
由…排出
easily because of its diuretic effects.

問41　どうして、ビールは何杯も飲めるの？

答 ①**大ジョッキ**のビールを3杯くらい平気（輕鬆）で飲める人でも、**同じ時間で同じ量の水を飲**

むことは容易ではありません（輕易）。②なぜでしょうか？

③水を飲むと胃に留（とど）まった後に、**十二指腸**を経て**小腸**、**大腸**とゆっくり送られ、大腸で**吸**

収されます。④短時間に**大量**の水を飲めないわけは、水を飲むスピードに、水を十二指腸に

送るスピードが**追**いつかず、胃をいっぱいにしてしまうからです。

⑤ビールには**エチルアルコール**が含まれています。⑥アルコールは**粘膜（黏膜）**などから体内に**吸収**

されやすい性質（特性）があります。⑦このアルコールの作用（作用）により、水が吸収（吸收）されない胃や十二指

腸で、ビールの20～30パーセント程度（程度）が吸収されると考えられています。

⑧吸収のよさに加えて、ビールには2つの**利尿作用（利尿的作用）**があります。⑨一つは、アルコールが**尿**を

抑制（抑制）する**ホルモン**の**分泌（分泌）**を抑える作用です。⑩もう一つは、ビールに含まれている**カリウム**

の利尿作用です。⑪吸収のよさと、利尿作用によるすみやかな排出（排出）により、ビールは短時間（短時間）

にたくさん飲むことができるのです。

問41　為什麼啤酒能夠一杯接著一杯地喝？

答 ①也許你可以輕鬆地喝下約3大杯的啤酒，但是卻無法輕易地在**相同時間**內喝下等量的水。
②這是為什麼呢？③當你喝水時，在胃裡短暫地停留後就會**流經十二指腸**、小腸、大腸，在大腸內被吸收。④短時間無法飲用**大量的水**的原因在於，水分運送到十二指腸的速度**無法跟上**你喝水的速度，胃裡面已經**充滿水分**的原故。
⑤啤酒之中含有**酒精**。⑥酒精有著能夠藉由**黏膜**組織等，輕易地被體內**吸收**的特性。⑦據說因為這酒精的作用，在水分無法被吸收的胃及十二指腸內，約有20～30%程度的啤酒被吸收。⑧除了容易被吸收之外，啤酒還有2個**利尿的作用**。⑨其中一個是，酒精有能夠控制抑制排尿的**荷爾蒙分泌**的作用。⑩另外一個是，啤酒中含有的**鉀**有著利尿的作用。⑪由於容易吸收以及因為利尿作用能夠迅速排出體外的原故，才能夠在短時間內大量地飲用啤酒。

Q42　Why does alcohol make people drunk?

①After you drink alcohol, your stomach and intestine
酒精　　　胃　　　腸
take it in. ②Then alcohol gets into the blood and goes to
吸收　　　　　　　　　　進入…　　　血液
the liver, where it is usually broken down. ③However, some alcohol
肝臟　　　　　　　　　　分解
which is not broken down and stays in the blood flows through
停留在…　　　　　　　流經…
blood vessels to the brain. ④This alcohol paralyzes the brain little
血管　　　大腦　　　　　　　　將…麻痺　　　　慢慢地
by little. ⑤This state is drunkenness.
酒醉
⑥As the blood-alcohol level goes up, the cerebral neocortex is
隨著… 血液中的酒精濃度　　　　　大腦新皮質
paralyzed and then the paralysis spreads to the cerebral limbic
麻痺　　　擴散　　　大腦邊緣系統
system, the cerebellum, the hippocampus, and finally to the
小腦　　　海馬迴　　　　　　　　　　延髓
medulla oblongata. ⑦When only the cerebral neocortex is
paralyzed, you are slightly drunk and flush a little. ⑧When the
微醺　　　　　　　臉頰微微泛紅
cerebellum is paralyzed, you become unsteady on your feet and
腳步凌亂
can not walk straight. ⑨You sometimes feel like vomiting in this
直線　　　　　　　感覺想吐
case. ⑩When the hippocampus, which is the center of the memory,
記憶中樞
is paralyzed, you can not speak well and you have a poor memory.
記憶力下降
⑪Finally, when the medulla oblongata is paralyzed, you can not
move your body and you are in serious danger.
在非常危險的狀態
⑫In the course of time, alcohol in the blood is broken down in the
隨著時間經過
liver. ⑬The brain recovers from the paralysis and you come to
由…中回復　　　　　　　甦醒過來
yourself as this breaking down continues and the blood-alcohol
繼續
level goes down. (see p.117)

問42　酒を飲むと、どうして酔っぱらうの？

答 ①酒を飲むと、**アルコール分はまず胃や腸で吸収されます。**②続いて**血液**に入り**肝臓**へ送られ、**分解されます。**③肝臓で分解されなかったアルコールは、血液の中に**残された**まま、**血管を通り、脳に達します。**④そして、脳を**麻痺**させていきます。⑤これが、酔っぱらった状態です。

⑥**血液中のアルコール濃度**が上がるにつれ、**大脳新皮質→大脳辺縁系→小脳→海馬→延髄**と麻痺が広がります。⑦**大脳新皮質が麻痺している状態は、顔がほんのり赤くなるほろ酔い状態**です。⑧**小脳が麻痺すると足元がふらつき、まともに歩けなくなります。**⑨**吐き気を感じることもあります。**⑩**記憶中枢**である海馬まで麻痺が進むと、ろれつがまわらなくなり、記憶が定かでなくなります。⑪そして、最後に延髄が麻痺すると、体を動かすことができなくなり、**非常に危険な状態**に陥ります。

⑫**時間がたつにつれ、血液中のアルコールは肝臓で分解されていきます。**⑬分解が**進むと血**液中のアルコール濃度が下がり、脳が麻痺**から回復して**酔いがさめます。

問42　喝了酒，為什麼會酒醉？

答 ①喝酒的時候，**酒精成分首先經由胃、腸而被吸收。**②接著進入血液中被送達了**肝臟、在這裡被分解。**③然而，一部分尚未在肝臟被分解的酒精殘留在血液之中，**通過血管、抵達大腦。**④隨後漸漸地麻痺大腦。⑤這就是所謂酒醉的狀態。

⑥隨著血液中的酒精濃度增加，麻痺現象會由大腦新皮質→大腦邊緣系統→小腦→海馬迴→延髓擴散。⑦大腦新皮質麻痺的狀態時，臉頰會微微地泛紅並呈現微醺的狀態。⑧小腦麻痺的時候，**腳步會凌亂、無法正常行走。**⑨有時會發生噁吐感。⑩當麻痺狀態前進到身為**記憶中樞**的海馬迴時，會產生口齒不清、記憶力下降的狀態。⑪接下來，當麻痺到最後的延髓時，身體會無法動彈、陷入非常危險的狀態之中。⑫隨著**時間**經過，血液中的酒精會由肝臟漸漸地分解。⑬隨著分解的**進行**，血液中的酒精濃度會下降，大腦由麻痺中**恢復**就會漸漸地酒醒了。

Human beings and chimpanzees separated from the common ancestor.

ヒトとサルは同じ祖先から枝分かれした
人類與猴子是由共同的祖先分支而來

Cercopithecidae
オナガザル科
獼猴科

Atelidae
クモザル科
蜘蛛猴科

Hylobatidae
テナガザル科
長臂猿科

Hominidae
ヒト科
人科

Spider monkey
クモザル
蜘蛛猴

Japanese monkey
ニホンザル
日本獼猴

Siamang
フクロテナガザル
合趾猴

Gibbon
テナガザル
長臂猿

Orangutan
オランウータン
紅毛猩猩

Gorilla
ゴリラ
大猩猩

Human
ヒト
人

Chimpanzee
チンパンジー
黑猩猩

Nakalipithecus
ナカリピテクス
仲山納卡里猿

Ouranopithecus
オウラノピテクス
歐蘭猿

5 million years ago
500万年前
500 萬年前

25 million years ago
2500万年前
2500 萬年前

40 million years ago
4000万年前
4000 萬年前

When you drink alcohol, you feel drunk more as your brain becomes paralyzed.

酒を飲むと、脳の麻痺が進むにつれて酔いが増す
當你飲酒的時候，隨著大腦麻痺、醉酒的程度會增加

Slight drunkenness
ほろ酔い状態
微醺的狀態

The cerebral cortex is paralyzed.
大脳皮質が麻痺
大腦皮質麻痺

You flush a little.
顔がほんのり赤くなる
臉頰微微泛紅

Tipsy stagger
千鳥足状態
微醉的狀態

The cerebellum is paralyzed.
小脳が麻痺する
小腦麻痺

You become unsteady on your feet and can not walk straight.
足元がふらつき、まともに歩けない
腳步凌亂，無法正常行走

Heavy drunkenness
酩酊状態
酩酊狀態

The hippocampus is paralyzed.
海馬が麻痺する
海馬迴麻痺

You can not speak well and you have a poor memory.
ろれつがまわらず、記憶が定かでない
口齒不清，記憶力下降

Coma
昏睡状態
昏睡狀態

The medulla oblongata is paralyzed.
延髄が麻痺する
延髓麻痺

You can not move your body.
体を動かすことができない
身體無法動彈

Q43　Why are feces brown?

①What you eat goes from the mouth to the small
　　　食物　　　　　　　　　　　　　　　　　　　　小腸
intestine by way of the gullet, the stomach, and the
　　　　　　　經由…　　　食道　　　　胃
duodenum. ②Nutrients are taken in mainly at the small intestine,
十二指腸　　　營養成分　　吸收
and the rest such as food fiber is sent to the large intestine with
　　　殘渣　　　　　食物纖維　　　　　　　大腸
water. ③At the large intestine, water is taken in and the waste is
　　　　　　　　　　　　　　　　　　　　　　　　　　排泄物
excreted as feces.
被排出　　作為糞便

④The coloring matter of feces is bilirubin, which is in bile made in
　　色素　　　　　　　　　　　　膽紅素　　　　　　　　膽汁
the liver. ⑤Bile is made in the liver and goes through the cholecyst
肝臟　　　　　　　　　　　　　　　　　　經由…　　　膽囊
and the bile duct to the duodenum, where what you eat is mixed
　　　膽管
with bile.

⑥Bile acid is found in bile as well as bilirubin. ⑦Bile acid makes it
膽汁酸　　　　　　　　和…一樣
easier to digest or take in fat. ⑧Bilirubin is made from broken-
　　　消化　　　　　　脂肪　　　　　　　　　　　　　　分解
down hemoglobins of old blood, which is a kind of the waste.
　　　血紅素
⑨Bilirubin is also called bile pigment, and it is yellow. ⑩The
　　　　　　　　　　　膽汁色素
Brown color of stercobilinogen made from
　　　　　　　黃膽元
bilirubin by intestinal bacteria and the yellow
　　　　　腸內細菌
color of the remainder of bilirubin are responsible
　　　剩餘的…　　　　　　　…的原因
for the color of feces.

問43

ウンチはどうして、茶色なの？

答 ①食べたものは口から**食道**、**胃**、**十二指腸を経て小腸**に送られます。②**小腸**など
食道　　　　　十二指腸　　　　　　　　　　　　　　　小腸
で**栄養分**が**吸収され**、**繊維分**などの**カス**が水分と一緒に**大腸**に送られます。③そし
営養成分　　　　　食物繊維　　　　水分　一同
て、大腸で水分が吸収されたものが、**ウンチ**となり、**肛門**から**排出される**のです。
肛門

④ウンチの色のもととなる**色素**は、**肝臓**でつくられる**胆汁**に含まれている**ビリルビン**
色素　　　　　　　　　　　膽汁
です。⑤胆汁は肝臓でつくられ、**胆嚢**、**胆管**を経て十二指腸に送られ、そこで食べた
膽嚢　　膽管
ものと混じります。

⑥胆汁には、ビリルビンのほかに**胆汁酸**が含まれています。⑦胆汁酸には、**脂肪を消化**
膽汁酸　　　　　　　　　　　　　　脂肪　消化
・吸収しやすくする働きがあります。⑧ビリルビンは、古くなった血液の**ヘモグロビ**
ンが**分解された**もので、**一種**の**排泄物**です。⑨**胆汁色素**とも呼ばれるもので、黄色を
一種　排泄物
しています。⑩このビリルビンから**腸内の細菌**によってつくられる**ステルコビリノー**
腸内　細菌
ゲンの**褐色**と、残りのビリルビンの**黄色**がウンチの色となるのです。
褐色　　　　　　　　　　　　　　黄色

問43

排泄物為什麼是咖啡色的呢？

答 ①**食物從口進入**，經由**食道、胃、十二指腸**最後送達了**小腸**。②**營養成分**在小腸等**被吸收**，**食物纖維**等的**殘渣**則與水分一同被運送到**大腸**。③接著，水分在大腸內被吸收，變成了排泄物，經由**肛門排出體外**。④形成排泄物顏色的**色素**，是肝臟所製造的膽汁當中含有的**膽紅素**。⑤膽汁在肝臟被製造，經由**膽囊、膽管**而被運送到十二指腸，在這裡與食物混合。⑥膽汁當中，除了膽紅素外還含有**膽汁酸**。⑦膽汁酸有促進**脂肪消化**、增加吸收的作用。⑧膽紅素是由老化血液中的**血紅素分解**而來，屬於排泄物的一種。⑨**膽紅色**又被稱之為**膽汁色素**，顏色呈黃色。⑩膽紅素經由**腸內細菌**製造出褐色的**糞膽元**，加上**剩餘的膽紅素**呈現的黃色，就是形成排泄物顏色的原因。

Q 44   Why do we dream?

① It is said that human beings dream mainly during
人類　　　　　　　　作夢　　　主要在…之間

REM sleep (rapid eye movement sleep), or light sleep.
REM睡眠　　（快速動眼期）　　　　　　　　　淺眠

② There are many unproved theories about why we dream, but a
　　　　　　　　　假說　　　　　　　為什麼會作夢

lot of questions have not been answered yet.
　　　　　　　還無法解答

③ The most believable theory is that dreams have something to do
　最有力的理論　　　　　　　　　　　　　　與…有關係

with a brain wave called the PGO wave (Ponto-geniculo-occipital
　　　腦波　　　　　　　PGO波

wave). ④ It is known that the brain sends out PGO waves during
　　　　　　　　　　　　　　　　釋放…

REM sleep. ⑤ There is a part of the brain which controls vision in
　　　　　　　　　　　　　　　　　　　　　　控制…　　視覺

the cerebral cortex. ⑥ A theory says that this part of the brain,
大腦皮質

stimulated by PGO waves, calls up broken visual pictures from
被…刺激　　　　　　　　　喚起…　片段的　視覺影像

memories.
記憶

⑦ PGO waves are sent out not only from the brain of human
　　　　　　　　　　　　　不僅…

beings but also from that of most other mammals during sleep.
　　　而且…　　　　　　　　　其他的哺乳類

⑧ If this theory is true, cats and

dogs are very likely to dream as
　　　有非常高的可能性…

well as human beings.
和…一樣

人はなぜ、夢を見るの？

答 ①**人間**が夢を見るのは、主に眠りが浅い**レム睡眠**中だといわれています。②その
人類　　　　　　　　　　　　　　　　　　睡眠

メカニズムについては、いくつかの**説**がありますが、はっきりとしたことはまだ**解明**
明瞭

されていません。

③なかでも有力なのは、**脳波の一種であるPGO波と関係している**という説です。④レム
脳波　　　　　　　　　　　関係

睡眠中には、PGO波が出ることが知られています。⑤**大脳皮質**には、**視覚を司る部分**
睡眠中　　　　　　　　　　　　　　　　　　　　　　　　　　　　視覺　掌管

があります。⑥レム睡眠中にPGO波がこの部分**を刺激する**ことにより、**記憶の中から**
刺激

断片的に視覚的なイメージが呼び起こされたものだとする説です。
片段

⑦PGO波は人間だけでなく、**ほとんどの哺乳類**が睡眠中に出すことが知られていま
哺乳類

す。⑧この説が正しいとすると、人間だけではなく犬も猫も夢を見ている**可能性が高**
可能性

くなります。

 人類為什麼會作夢？

答 ①據說**人類**作夢的時間，主要是在REM睡眠（快速動眼期）或是淺眠期間。②關於作夢的
機制有許多**假說**，但是正確的原因**目前還未完全明瞭**。③其中最有力的一種理論是，作夢
與一種稱之為PGO波的腦波有著密切的關係。④我們已知，在REM睡眠中會釋放出PGO波。⑤腦
部中的**大腦皮質**，負責掌管視覺的部位。⑥有此一說是，由於在REM睡眠中所釋放的PGO波會刺
激這個部位，從而**喚起記憶中片段的視覺影像**。
⑦PGO波**不只人類**，幾乎所有的哺乳類在睡眠中都會釋放。⑧如果這個理論正確的話，貓或狗會
作夢的**可能性**也非常高。

Q45　Why do we have fever, cough, and a sore throat when we catch cold?

①A cold is caused by disease agents such as germs and viruses. ②When these disease agents enter the human body, the body tries to protect its cells from the attack by them. ③This is what we call immune reaction.

④The first immune reaction is inflammation, which is the cause of coughs, sore throats, and so on. ⑤Today, it is made known that what causes inflammation is a protein generally called inflammatory cytokine. ⑥It is made by white blood cells named macrophages, which play a part in fighting against the disease agents. ⑦Cytokines attract white blood cells by inflammation and coagulate blood and stop the disease agents from spreading to other parts in order to keep the sickness from getting worse.

⑧Cytokine plays a role in fever, one of the inflammation reactions. ⑨When cytokines get to the brain, the temperature control center orders the body to raise its temperature. ⑩Rise in body temperature keeps disease agents from increasing and makes white blood cells work actively, so the sickness is likely to be cured.

問 45　風邪をひくと、なぜ熱やせきが出たり、のどが痛くなったりするの？

答 ①風邪の原因となるのは、**細菌**やウィルスなどの**病原体**です。②病原体が体内に入る
感冒　　原因　　　　　　　　細菌　　　　　　　　病原體　　　　　　　身體

と、**人間の体**はこれらの**攻撃**から**細胞を守ろう**とします。③いわゆる**免疫反応**です。
　　　　　　　　　　　　攻擊　　　　　　　　　　　　　免疫反應

④免疫反応として**最初**に起こるのが、**せきやのどの痛み**などの**原因**となる**炎症**です。⑤炎症
　　　　　　　　最初　　　　　　　　　　　　　　　　　　　　　　炎症

を引き起こすのは、炎症性サイトカインと**総称される**タンパク質であることが明らかになっ
　　　　　　　　　　　　　　　　　通稱

ています。⑥これは、病原体と戦う役割を持った**マクロファージ**と呼ばれる**白血球**によって
　　　　　　　　　　　　　　負責　　　　　　　　　　　　　　　　　　白血球

つくり出されるものです。⑦炎症により、白血球**を呼び寄せ**たり、**血液を凝固させて**病原体
　　　　　　　　　　　　　　　　　　　　　　　　　血液　凝固

がほかの場所**に広がるの**を抑えたりして、病気の**悪化**を**防ぎ**ます。
　　　　　　　　　　　　　　　　　　疾病　　惡化

⑧**発熱**という炎症反応にもサイトカインがかかわっています。⑨サイトカインが脳に達する
發燒

と、**体温調節中枢**が体温**を上げる**指令を出します。⑩体温**が上がる**ことにより、病原菌の
體溫調節中樞　　　　　　　　　　　　　　　　　　　　　　　　　　　　病原體

増殖が抑えられるとともに、白血球が**活発に働き**、病気が**治り**やすくなるからです。
增殖　　　　　　　　　　白血球　活躍　　　　　疾病

問 45　感冒的時候，為什麼會發燒、咳嗽、喉嚨痛呢？

答 ①感冒**是由細菌或病毒**等的**病原體**所引起的。②當病原體侵入身體時，人類的**身體**必須由這些**攻擊**中守護細胞。③這就是我們所稱的免疫反應。

④由免疫反應所引起的最初的**炎症**，是造成咳嗽或喉嚨痛等等的原因。⑤現今我們知道，炎症的誘發是由一種**通稱**為炎性細胞因子的蛋白質所引起。⑥這些蛋白質是由負責與病原體作戰，**被稱為巨噬細胞**的白血球所形成。⑦因為炎症而**招來白血球**，它能夠讓血液凝固並且抑制病原體擴散到其他的位置，**防止疾病惡化**。

⑧**發燒**的炎症反應也跟細胞因子有關。⑨當細胞因子抵達**腦部**，**體溫調節中樞**會下達體溫升高的命令。⑩由於**體溫升高**，病原體的**增殖**跟著被抑制、白血球**開始活躍**，所以疾病就容易被治癒。

Q46 Why does a headache go away with medicine taken through the mouth?

①It is easy to understand that stomach medicine can
胃藥
cure stomachaches because it melts in the stomach.
治療 胃痛   溶解
②Then, how does medicine work when we take it for headaches?
那麼   頭痛
③Medicine for headache such as aspirin melts in the stomach and
阿斯匹靈
enters the small intestine. ④After that, it gets into the blood and
進入   小腸   進入   血液
goes to all parts of the body by way of the liver. ⑤Medicine for
經由…   肝臟
headaches gets to the head, the painful part, through blood vessels
患部   通過…   血管
15 to 30 minutes after taking the medicine.
15~30分鐘後
⑥One of the causative agents of headache and period pain is
致病媒介   生理痛
prostaglandin. ⑦It produces pain by stimulating the nerves.
前列腺素   引發疼痛   刺激   神經
⑧Over-the-counter medicine for headaches such as aspirin works
市售藥
to keep the prostaglandin from causing pain. ⑨This effect eases
阻止…   引起…   緩和
pain.

⑩To sum up, medicine for headaches taken through the mouth
也就是說   經由口服
enters the small intestine and is carried to all parts of the body
被運送到…
through blood vessels, and then it removes the cause of pain and
去除…
the headache goes away.
治癒頭痛

問46　口から飲んだ薬で、なぜ頭痛が治るの？

答 ①**胃薬**を飲むと、薬が胃で溶けるため**胃痛**が**治る**のはわかります。②では、**頭が**

痛いとき飲んだ薬はどのように効くのでしょうか。

③**アスピリン**に代表される**頭痛薬**は、胃で溶け、**小腸で吸収されます**。④その後、**血液**

に入り、**肝臓を経て全身**に運ばれます。⑤頭痛薬は、**服用後15〜30分**程度で血流に乗

って**患部**である頭部に到達します。

⑥頭痛や**生理痛**の原因の一つとして、**プロスタグランジン**という**物質**があげられま

す。⑦これが**神経**を**刺激する**ことにより、**痛みが起こります**。⑧アスピリンなど**市販の**

頭痛薬には、プロスタグランジンが痛み**を引き起こす**ことを**妨げる**作用があります。

⑨この作用が、痛みを**やわらげます**。

⑩つまり、**口から服用した**頭痛薬が小腸から吸収され、血液によって患部**に運ばれ**、

これが痛みの原因**を取り除く**ため、**頭痛が治る**のです。

問46　為什麼口服藥可以治療頭痛？

答 ①我們可以理解，服用**胃藥**時，因為藥在胃裡溶解而能**治療胃痛**。②那麼，頭痛的時後服用口服藥，為什麼可以產生效果呢？③代表性的頭痛藥如**阿斯匹靈**，在胃裡溶解、在小腸被吸收。④之後進入血液，經由肝臟運送到全身。⑤頭痛藥在服用後約**15〜30分鐘**後，會隨著血液抵達**身為患部**的頭部。

⑥頭痛或**生理痛**的原因之一，是由被稱為**前列腺素的物質**所引起。⑦由於它**刺激神經而引發疼痛**。⑧像阿斯匹靈等**市售**的頭痛藥，含有**防止**前列腺素**引發疼痛的作用**。⑨因為這個作用，能夠緩和疼痛。

⑩**也就是說**，經由口服的頭痛藥被小腸吸收，透過血液**運送**到患部，並且**去除**了疼痛的原因，所以**治癒了頭痛**。

Q47 Why doesn't the stomach wall dissolve in stomach acid?

① Strong acid liquid and an enzyme that breaks down protein named pepsin come out in the stomach. ② The strong acid liquid and pepsin digest substances so strongly that they can break down meat. ③ However, they do not digest the stomach itself, a muscular organ. ④ Why not?

⑤ The stomach wall is covered with mucous membranes which produce mucus. ⑥ Mucus is made up mainly of mucin, which is the cause of stickiness in fishskin, taro, fermented soybeans, and so on.

⑦ A mucous gel layer a few millimeters thick is on the surface of mucous membranes. ⑧ The gel is something like konnyaku jelly or gelatin. ⑨ When mucous membranes are broken down by strong acid liquid or pepsin on the surface of a gel layer, new mucous membranes take the place of the old ones. ⑩ The reason why the stomach wall does not dissolve in the strong acid liquid or the digestive enzyme is that this gel layer protects the stomach.

問47　胃の壁はなぜ、胃酸で溶けないの？

答 ①胃の中には強い**酸性の液**とペプシンという**タンパク質**分解**酵素**があります。

②これらには**肉類**も分解してしまうほど、強力な消化作用があります。③しかし、**筋肉**ででできている**胃自体**は消化されることがありません。④これはなぜでしょうか？

⑤**胃の壁は粘膜**で覆われ、**粘液が出されています。**⑥粘液の主成分は**ムチン**と呼ばれるもので、**魚の皮、サトイモ、納豆**などのぬめり**のもと**となる物質です。

⑦粘膜の**表面**では、**粘液が厚さ数ミリメートルのゲル層**をつくっています。⑧ゲルというのは、**コンニャクやゼラチン**のような状態です。⑨ゲル層の表面で強い酸やペプシンにより粘液が**分解**されると、新しい粘液に置き換わるしくみが働いています。⑩胃の壁が強い酸や**消化酵素**の**影響**を受けないのは、このゲル層によって守られているからです。

問47　為什麼胃壁不會被胃酸溶解？

答 ①在胃裡含有強酸性的液體，以及稱為**胃液素**的蛋白質分解酵素。②這些液體，擁有連**肉類**都能夠分解的強力消化作用。③但是，它卻不會消化由**肌肉**所組成的胃本身。④這是為什麼呢？⑤**胃壁**覆蓋著**黏膜**，並會產生**黏液**。⑥黏液的主要成份稱之為**黏蛋白**，是由魚皮、芋頭、**納豆**等的**黏稠物**為基本所形成的物質。⑦黏膜的**表面**，由厚達數厘米的黏液形成了黏液層。⑧這黏液層的狀態有如**蒟蒻**或是**明膠**。⑨當黏液層表面的黏液因為強酸或胃液素被分解，新的黏液就會取代舊有的表層。⑩胃壁為什麼不會受到強酸或**消化酵素**的影響，就是因為有黏液層守護的原故。

Q48 Why does the head hurt after eating shaved ice?

① The headache you have after eating something cold is
頭痛　　　　　　　　　　　　　　　　　　　　一些冰涼的食物

called an ice cream headache. ② There seem to be two
冰淇淋頭痛　　　　　　　　　　　　　似乎

ideas about why you get an ice cream headache.
說法、臆測

③ One of them is an idea which has something to do with nerve
　　　　　　　　　　　　　　　　　　與…有關

messages. ④ When you eat something cold, information that it is
神經傳達　　　　　　　　　　　　　　　　情報

cold in the mouth is sent to the brain through the trifacial nerve.
口　　　　　被送到…　腦部　　　　　　　三叉神經

⑤ The stimulus of ice cream and ice, which are very cold, are so
刺激　　　　　　　　　　　　　　　　　　　　如此…以致於…

strong that information about nerves is confused and sent to the
　　　　　　　　　　　　　　　　　混亂

brain as information that the head hurts.
疼痛

⑥ The other idea is that something cold causes inflammation of
　　　　　　　　　　　　　　　　　引發…　炎症

the blood vessels. ⑦ A reaction to try to increase blood flow is
血管　　　　　　　反應　　　　　　增加　血流

caused to warm the inside of the mouth which is cooled down by
溫熱…

the ice. ⑧ As a result, this idea says the

blood vessels in the brain become inflamed
引起發炎

temporarily because the blood vessels in
暫時的

the brain suddenly swell.
忽然地　變粗

問48 かき氷を食べると、なぜ頭痛がするの？

答 ①冷たいものを食べたあとに起こる**頭痛**は、「**アイスクリーム頭痛**」と呼ばれて
います。②なぜ、アイスクリーム頭痛が起こるのかについては、二つの説が**あるよう**

です。

③一つは、**神経**による**伝達**に関係があるとする説です。④冷たいものを食べると、「口
の中が冷たい」という**情報**が**三叉神経**を経て脳に**伝わります**。⑤アイスクリームや氷
は非常に冷たいため、その強烈な**刺激**によって神経の情報が**混乱し**、「頭が**痛い**」と
いう情報として伝わってしまう、というものです。

⑥もう一つは、冷たいものが**血管**の**炎症**を引き起こすとする説です。⑦冷たくなった口
の中を温めるため、**血流**を増やそうとする**反応**が起こります。⑧そこで、頭の血管が
急激に太くなるため、頭の血管が**一時的に炎症を起こす**、という説です。

問48 為什麼吃冰之後會產生頭痛？

答 ①吃下太**冰涼**的食物之後所引起的**頭痛**，稱之為「**冰淇淋頭痛**」。②關於為什麼會引起冰
淇淋頭痛，**似乎有以下的兩種說法**。③其中一個說法是，與**神經傳達**有關。④當冰涼的食
物進入口中，「口中很冰涼」的**情報**就會經由三叉**神經傳達到腦部**。⑤由於冰淇淋或冰塊相當冰
涼，因為它的強烈**刺激**而引起情報**錯亂**，傳送出「**頭痛**」的情報。
⑥另外一個說法是，冰涼的食物會導致**血管引發炎症**。⑦這是由於要將變得冰涼的口腔變得**溫
熱**，**血流增加**所引起的生理反應。⑧於是，腦中的血管突然間變粗，導致腦中的血管**引起暫時的
炎症**，有這樣的說法。

You dream mainly during REM sleep (rapid eye movement sleep).

夢は、レム睡眠時に見ることが多い
作夢主要是在REM睡眠期間

During REM sleep
レム睡眠の状態
REM睡眠的狀態

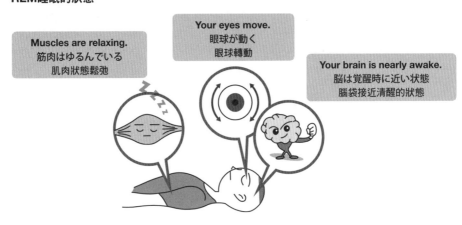

Muscles are relaxing.
筋肉はゆるんでいる
肌肉狀態鬆弛

Your eyes move.
眼球が動く
眼球轉動

Your brain is nearly awake.
脳は覚醒時に近い状態
腦袋接近清醒的狀態

How you dream
夢を見るしくみ
如何作夢

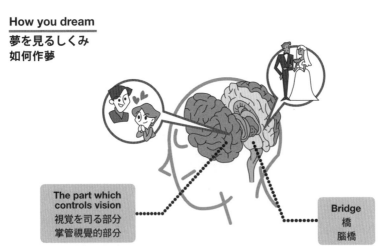

The part which controls vision
視覚を司る部分
掌管視覺的部分

Bridge
橋
腦橋

The PGO waves the "bridge" in the brain stem gives off stimulate the cerebral cortex and call up visual pictures of the dream.
脳幹にある「橋」から出るPGO波が大脳皮質を刺激し、夢の視覚イメージが生まれる
由位於腦幹中的腦橋生成的PGO波刺激大腦皮質，產生了夢的視覺影像。

This is how you have headache after eating shaved ice.

かき氷を食べると頭痛がするしくみは？
吃了冰涼的食物，會產生頭痛的機制是？

The strong stimulus causes information in the brain confused.

① 強い刺激で、脳が混乱する
①因為強烈的刺激，使大腦產生混亂

When information that it is cold is sent to the trifacial nerve, the brain is confused and thinks of it as information that it hurts.

冷たいという強い刺激が三叉神経に伝わると、脳が混乱し、「痛い」という情報になる

冰涼的強烈刺激傳達到三叉神經，使大腦混亂、產生「疼痛」的情報。

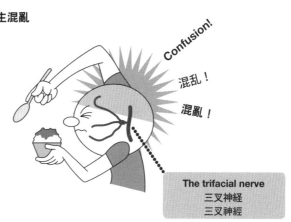

Confusion!

混乱！

混亂！

The trifacial nerve
三叉神経
三叉神經

The brain vessels swell.

② 脳の血管が拡張する
②腦部的血管擴張

Blood flow is increased to warm the inside of the mouth and the brain vessels swell suddenly, so you feel pain.

血流を増やして冷えた口の中を温めるため、頭の血管が急に太くなり痛みを感じる

由於要溫熱變得冰涼的口腔，血液流動增加、腦部的血管突然間變粗而感覺到疼痛。

The vessels swell and stimulate the nerves.
血管が太くなり、神経を刺激する
血管變粗、刺激到神經

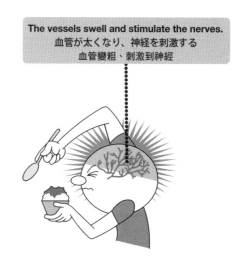

Q49   Why do you have white hair?

①There is a hair papilla at the base of a hair. ②Hair
　　　　　　毛乳頭　　　　　　　　　毛髮的根部　　　　　毛母細胞

matrix cells, which grow hairs, are at the top of the hair

papilla. ③Nutrients are sent to hair matrix cells through capillary
　　　　　養分　　　　　　　　　　　　　　　　　　　毛細血管

vessels and the hair matrix cells divide to grow a hair.
　　　　　　　　　　　　　　　　　　分裂

④There are pigment cells called melanocyte around the hair
　　　　　　色素細胞　　　　　　　黑色素細胞

papilla. ⑤Melanin pigment which is produced in the melanocyte
　　　　　黑色素

goes into the hair, so clear hair becomes black.
　　　　　　　　　透明的

⑥Hairs repeat the cycle of the anagen phase, catagen phase, and
　　　反覆地　　週期　　　　　生長期　　　　　　退化期

telogen phase. ⑦The anagen phase lasts for three to six years and
休止期　　　　　　　　　　　　　持續

hair grows during this phase. ⑧The catagen phase lasts for a few
　　　　　　　　　　階段

weeks and hair stop growing during this phase. ⑨The telogen

phase lasts for three to four months, during which the hair falls
　　　　　　　　　　　　　　　　　　　　　　　　　脫落

out and the hair matrix cells get ready to grow new hair.
　　　　　　　　　　　　　準備

⑩In the telogen phase, you lose melanocyte with your hair. ⑪If

you are young and have energy, melanocyte is produced to grow
　　　　　　　　　　　精力

new hair and is placed around the hair papilla. ⑫However, you
　　　　　　　被分配在…

have white hair when the melanocyte is not properly placed and

the hair is not filled with it because of sickness, aging, stress, and
　　　　　　　　充滿　　　　　　　　　　　　　老化　壓力

so on.

問49　なぜ、白髪になるの？

答 ①髪の毛のつけ根の部分には、毛乳頭があります。②その先端に、毛髪をつくる毛
　　　　　　　　　　　毛乳頭　　　　　　　　　　　　　前端　　毛髪
母細胞があります。③毛細血管により養分が運ばれ、毛母細胞が細胞分裂をして毛髪をつ
毛母細胞　　　　　　毛細血管　　　　養分
くります。

④毛乳頭の周辺には、メラノサイトという色素細胞があります。⑤メラノサイトでつくられた
　　　　　周圍　　　　　　　　　　　　色素細胞
メラニン色素が毛髪に入るため、透明だった毛髪が黒くなります。
　　　　　　　　　　　　　　　　透明

⑥髪の毛は成長期→退行期→休止期というサイクルを繰り返しています。⑦成長期は3〜6
　　　　生長期　退化期　休止期
年間で、この間に髪の毛が伸びます。⑧退行期は2〜3週間で、髪の毛の成長が止まります。

⑨休止期は3〜4ヵ月で、脱毛の後、毛母細胞が次の発毛の準備をします。
　　　　　　　　　　　脱毛　　　　　　　　　　　　生髪　準備
⑩休止期に脱毛するときには、毛髪とともにメラノサイトが失われます。⑪若くて元気なら、

再び発毛するときにメラノサイトが作られ、毛乳頭の周辺に配置されます。⑫しかし、病気
　　　　　　　　　　　　　　　　　　　　　周圍　分佈　　　　　　　　　　　　　生病
や加齢、ストレスなどによりメラノサイトがうまく配置されなくなると、髪の毛にメラニン色
年齡增長　　　　　　　　　　　　　　　　　　　　　　　　　　　　　　　　　　　　色素
素がいき渡らなくなるので、白髪になります。
　　　　　　　　　　　　　白髮

問49　為什麼頭髮會變白？

答 ①毛髮的根部有毛乳頭。②在它的前端有著製造毛髮的毛母細胞。③養分經由毛細血管運送到這裡，毛母細胞借由細胞分裂而製造出毛髮。④在毛乳頭的周圍，有著稱為黑色素細胞的色素細胞。⑤由於黑色素細胞製造的黑色素進入了毛髮當中，使得透明的毛髮變成了黑色。⑥毛髮反覆地進行生長期→退化期→休止期這樣的循環。⑦生長期為3〜6年，這個期間毛髮會生長。⑧退化期為2〜3週，毛髮的成長停止。⑨休止期為3〜4個月，毛髮掉了之後，毛母細胞會準備下一次的生髮。⑩在休止期脫毛的時候，黑色素細胞會隨著毛髮流失。⑪當年輕力壯的時候，毛髮再生時也會製造黑色素細胞，並且分佈在毛乳頭的周圍。⑫但是，若是因為生病、年齡增長、壓力等因素而無法製造出黑色素細胞來分佈在其周圍，毛髮因為缺乏了黑色素，就會形成白髮。

Q50 Where do baby's feces go in the mother's body?

① A baby in its mother's body spends about 9 months
花費…
floating in amniotic fluid in her womb. ②The baby gets
羊水   子宮
nutrients and oxygen from the mother through the umbilical cord.
養分   氧氣   臍帶
③The baby in its mother's body doesn't eat anything, so it has no
                                        什麼東西   沒有糞便
feces. ④The baby drinks amniotic fluid through its mouth and
urinates back into amniotic fluid. ⑤During this period, the baby
排尿   進入   這段期間
keeps wastes, such as pieces of skin, hair, and so on, in its
保持 廢物   皮膚碎屑
intestine. ⑥The first feces the baby excretes after birth are called
自己的腸子   排泄   出生後
meconium. ⑦Meconium is the waste of amniotic fluid kept for 9
胎便
months and it has a dark color and little smell.
                    深色
⑧When the baby stays in its mother's body even after the expected
                                    甚至超過了…   預產期
date of its birth, the placenta doesn't work well and the baby has
              胎盤
discomfort and feces float in amnionic fluid. ⑨If the feces go into
感覺不適
the baby's lung through the throat, the baby sometimes has
          肺部   通過氣管
difficulty breathing and it gets into danger.
呼吸困難   發生危險
⑩If it takes too long for the baby to excrete meconium after birth,
      花費…為了…
the baby is likely to be jaundiced because the bilirubin, yellow
          容易   形成黃疸   膽紅素   黃色色素
coloring matter in the meconium gets into the blood again from
                                        進入
the intestine.

問50　お腹の中の赤ちゃんのウンチは、どうなっているの？

答 ①赤ちゃんはお母さんのお腹（なか）の中で、**子宮（しきゅう）の中の羊水（ようすい）**に浮かんで約9ヵ月間**過ごしま**
す。②赤ちゃんはお母さんから**へその緒を通じて栄養と酸素**をもらいます。③食べ物を食べ
るわけではないので**ウンチはしません**。④口から羊水を飲んで、**おしっこをまた羊水の中に**
戻します。⑤そのときに羊水に含まれる**皮膚や毛などのゴミ**を、**自分の腸の中にためておき**
ます。⑥**生まれてから初めて排泄（はいせつ）するウンチを胎便（たいべん）**といいます。⑦それは、9ヵ月の間にたま
った羊水のゴミなどで、**色が黒っぽくて臭（にお）い**はあまりありません。

⑧**出産予定日**が過ぎてしまうと、**胎盤の機能が低下**し赤ちゃんが苦しくなって、羊水の中で
胎便が出てしまうことがあります。⑨これが赤ちゃんの**気管から肺に入る**と、出産後に**呼吸**
障害を起こすことがあり、**危険**です。

⑩また、生まれてから胎便が出るまで**に時間がかかる**と、胎便に含まれる**黄色の色素のビリ**
ルビンが腸管から血中へ再吸収されるので、**黄疸になりやすくなります**。

問50　腹中胎兒的糞便，要如何運送到母體？

答 ①胎兒在母親的肚子裡，會浮在子宮當中的羊水、度過約9個月的時間。②胎兒透過臍帶，從母親這裡獲得**養分與氧氣**。③由於沒有進食，理所當然地不會製造**糞便**。④羊水經由口中飲入，**小便也再次排放回羊水中**。⑤在這段期間，羊水中包含的**皮膚碎屑，毛髮等廢物**，會堆**積在胎兒的大腸**。⑥出生後初次排泄的糞便就稱之為**胎便**。⑦這是在九個月的期間，堆積在羊水中等等的廢物，**顏色呈深黑色、幾乎沒有臭味**。
⑧若是超過了**預產期**，**胎盤的功能退化**導致胎兒不適，胎便就會解到羊水之中。
⑨這些經由胎兒的**氣管進入肺部**，在出生後會**引起呼吸障礙，非常危險**。⑩此外，出生後若是**耗費太多時間解胎便**，胎便中包含的**黃色色素膽紅素**，會經由腸道再次吸收進入血液之中，**容易形成黃疸的現象**。

Q51 Why don't human beings breathe in water?

①Mammals 哺乳類 such as human beings 人類 breathe 呼吸 with the lungs 肺. ②There are many small bags named air sacs 肺泡 and they contain a lot of capillary vessels 毛細血管 are in them. ③Air taken in 被吸入的空氣 through breathing 呼吸 goes through the throat 氣管 and eventually 最後 reaches the air sacs. ④There, 在這裡 the air is taken in by blood flowing through 流經… the capillary vessels and, at the same time, carbon dioxide 二氧化碳 and water 水分 are exuded 排出.

⑤This system 機制 does not work in water. ⑥Imagine that 想像… you take in water instead of 取代 air. ⑦When air sacs are filled with 充滿… water, oxygen 氧氣 can not be exchanged for 與…交換 carbon dioxide in the blood. ⑧Furthermore, 此外 water goes into your blood when you take in 吸入 pure water 純水 and water in blood comes out 出來 when you take in 吸入 sea water, 海水 so the balance in 平衡 the blood is lost 失去 in either case. 在任何的場合 ⑨This causes death.

⑩The baby in mother's body can live in amniotic fluid 羊水 because it does not breathe with the lung but it gets oxygen 氧氣 from the mother and releases carbon dioxide to the mother through the placenta. 胎盤

問51　人はどうして、水の中で息ができないの？

答 ①**人間**をはじめとする**哺乳類**は肺で呼吸します。②肺の中には**肺胞**という無数の
小さな袋があり、**毛細血管**が集まっています。③**呼吸**することによって**取り込まれた**
空気は、**気管**を通り**最終的に**肺胞に送られます。④そこで、毛細血管を流れる血液に
酸素が取り込まれ、同時に**二酸化炭素**や**水分**が**排出されます**。
⑤このしくみは、水中では通用しません。⑥仮に空気**の代わりに**水を吸い込んだとしま
しょう。⑦肺胞が水で**満たされる**と、血液の**酸素**と二酸化炭素の**交換**ができなくなり
ます。⑧**さらに**、水の場合は浸透圧によって水が血液中に入り、**海水**の場合は逆に血
液中の水分が流出して、血液の**バランス**が崩れてしまいます。⑨これは人間にとって
死を意味します。
⑩胎児が**羊水**の中で生きていられるのは、肺で呼吸せず、**胎盤**を通して母親から**酸素**
をもらい、二酸化炭素を戻しているからです。

問51　人類在水中為什麼無法呼吸？

答 ①以**人類**為首的哺乳類是用肺來呼吸。②在肺中含有無數個稱為**肺泡**的小氣囊、且當中包
含著毛細血管。③藉由呼吸取得的空氣，通過氣管，**最終**被送達肺泡。④**在這裡**，毛細血
管中的血液攝取氧氣、同時間將二氧化碳與**水分**排出。
⑤這個**機制**在水中卻無法適用。⑥假設水**取代**空氣，被吸入肺中。⑦肺泡中**積滿**了水，血液中的
氧氣與二氧化碳**的交換**便無法進行。⑧接著，若是吸入水分，根據滲透壓原理，水分會進入血液
中；若是吸入**海水**，相反的血液中的水分就會流失，兩者血液的**平衡**均會喪失。⑨這對人類而言
就意味著死亡。
⑩胎兒在羊水之中可以生存，是因為不用肺部呼吸，而是藉由**胎盤**連接母體，取得氧氣、並將二
氧化碳送回母體的緣故。

Q52 How do you find a criminal by a fingerprint?

①There are no pores at the tip of a finger but there are
毛囊　　　　　　　　指尖

sweat glands. ② The open mouths of the sweat glands are
汗腺　　　　　　　　開口部　　　　　　　　　　　　凸起的

raised and they form a pattern of equal-spaced ridges like
　　　　　　　　形成　　圖樣　　　　　　等間隔的隆線

a contour map. ③This is the fingerprint.
等高線圖案　　　　　　　　指紋

④The pattern of a fingerprint never changes from the cradle to the
　　　　　　　　　　　　　　　　　　　　　從搖籃到墳墓（→一生）

grave. ⑤ Even if you chip off the skin of the tip of your finger, the
即使…　　　　削掉　　　　　　　　　　　　　　　　　　　　再生

renewed tip has the same pattern of fingerprint as before. ⑥ It is
　　　　　　　　　　相同

true that the finger grows thicker and the fingerprint grow bigger
可以確定的是…　　　　變粗

as people grow up but the pattern of the fingerprint never
隨著…

changes.

⑦There is no other person that has the same fingerprint as you in

the world. ⑧You can not tell the difference between identical twins
　　　　　　　　　　　區別　　　　　　　　　　同卵雙胞胎

by DNA test but you can tell by their fingerprints. ⑨ However, the
DNA鑑定

fingerprints of identical twins are not completely different. ⑩They
　　　　　　　　　　　　　　　　並非完全

have similar patterns of fingerprint because they have same genes
　　相似的　　　　　　　　　　　　　　　　　　　　　　遺傳因子

but there is difference enough to tell one from the other.
　　　　　　　　　　　　　區別雙方

問52　指紋でなぜ、犯人が見つかるの？

答 ①**指先**には**毛穴**はありませんが、**汗腺**があります。②汗腺の**開口部**は**隆起**し、**等高線**のように**一定間隔**の**隆線**という**模様**をつくります。③これが**指紋**です。

④指紋は、一生変わりません。⑤たとえ指先の**皮膚**を**はぎ取**っても、皮膚ができると同じ指紋になります。⑥子どもから大人に**成長**すると指が**太く**なり、指紋も大きくなりますが、パターンは変化しません。

⑦また、世の中に同じ指紋の人は**存在**しません。⑧**DNA鑑定**で**区別**できない**一卵性双生児**も、指紋鑑定なら特定することができるのです。⑨ただし、一卵性双生児の指紋はまったく異なっているわけでもありません。⑩**遺伝子**が同一であるために指紋も**似**たパターンですが、**指紋認証**で区別できるほどの違いはあります。

問52　為什麼利用指紋就可以找出犯人？

答 ①指尖雖然沒有毛囊，但是卻有著汗腺。②汗腺的開口部隆起，形成有如等高線一般，有著一定間隔、稱之為紋線的圖樣。③這就稱之為指紋。

④指紋一生都不會改變。⑤假設剝掉指尖的皮膚，當皮膚生成之後會形成相同的指紋。⑥由兒童成長為大人時，雖然手指頭變粗、指紋變大，但是圖樣並不會改變。

⑦此外，世界上並不存在擁有相同指紋的人。⑧即使經由DNA鑑定也無法區別的同卵雙胞胎，經由指紋辨識也能判別出不同。⑨但是，同卵雙胞胎的指紋並非完全不同。⑩由於有相同的遺傳因子而有相似的指紋，但是經由指紋辨識還是有足夠的程度判別出不同。

Q53

Why don't we see things double though we have two eyes?

①We have two eyes, but we do not see things double.
看見兩個物體
②Why don't we see double?

③There is a crystal lens in our eye. ④Light passing through the
水晶體   穿過…
crystal lens strikes the retina at the back of the eyeball. ⑤The
到達   視網膜   在後面   眼球   影像
image the retina receives is changed into signals by the nerves in
接收   轉換為   信號   視神經
the eyes lining up on the retina, and the signals travel to the
排列在…   傳送到   大腦
cerebrum as two images on the right and the left.
                              左右兩邊
⑥The two images which travel as the signals through the nerve to
當作
the cerebrum from the right eye and the left eye are overlapped by
與…部分
the following three functions of the cerebrum. ⑦The first one is
以下的   機制
simultaneous perception, which is a function allowing us to see
同時視
two different images at the same time. ⑧The second is the fusion,
融像
which is a function which allows two images to overlap and make
將…變成
them into one. ⑨The third is binocular vision, which is a function
立體視
which creates a feeling of distance and depth by small differences
遠近感與立體感
between the two images on the right and the left.

⑩Images produced on the retinas of the two eyeballs come together
由…作成   合在一起
because of these three functions of the cerebrum. ⑪That is why we
因為如此
can see things in 3D. (see p.144)
立體的

問 53　眼は二つあるのに、なぜものが一つに見えるの？

答 ① 私たちには眼が二つありますが、見えている像は一つだけです。② これは、なぜでしょうか？

③ 眼には**水晶体**というレンズがあります。④ 水晶体を**通過**した光は、眼の一番**奥**にある**網膜**
水晶體　　　　　　　　　　　　　　　　　　穿過　　　　　　　　　　　　　　　　網膜（もうまく）
に映し出されます。⑤ 映し出された**像**は、両眼の網膜に並ぶ**視神経**によって信号に**変換さ**
うつ　　　　　　　　　　　　　　　両眼　　　　　視神經　　　　　信號　　轉變
れ、左右二つの像として**大脳**に伝えられます。
　　　　　　　　　　　　　　大腦

⑥ 大脳に備わっている**次の三つの働き**によって、右眼と左眼から神経を通じて信号**として送**
　　　　　　　　　　　　　　　　　　　　右眼　　左眼　　神經
られてきた二つの像が**重ね合わせられます**。⑦ まず、「**同時視**」は、二つの異なった像を同
　　　　　　　　　　　　　　　　　　　　　　　　　　同時視
時に見る働きです。⑧ 次に、「**融像**」は二つの像を重ね合わせ、一つの像にする働きです。⑨
　　　　　　　　　　　　融像
そして、「**立体視**」は、左右の像の微妙な違いから、像に**遠近感や立体感**を与えます。
　　　　　立體視　　　左右　　些微　　　　　　　　遠近感　立體感
⑩ **大脳**が持つ、これら三つの働きにより、二つの眼に映し出された像は**一つになります**。⑪ そ
大腦
して、私たちは**立体的**にものを見ることができるのです。
　　　　　　　立體的

問 53　為什麼我們有兩隻眼睛，卻只看得見一個物體呢？

答 ① 雖然我們有兩隻眼睛，但是看見的影像卻只有一個。② 這是為什麼呢？
③ 眼睛裡有稱之為「**水晶體**」的鏡頭。④ **穿過**水晶體的光線，映照在位於眼睛最**深處的視**
網膜上。⑤ 映照出來的**影像**，經由排列在兩眼視網膜上的**視神經轉變為信號**，將左右兩個影像傳
送給大腦。
⑥ 而大腦就根據**以下**的三個機制，將右眼以及左眼透過神經所傳達的信號，兩者的影像**合而為**
一。⑦ 首先是「**同時視**」，在同一時間看見兩個不同影像的機能。
⑧ 接下來是「**融像**」，將兩個影像重疊、合成一**個**影像的**機能**。⑨ 最後是「**立體視**」，借由左右
影像的些微差異，賦予影像**遠近感與立體感**。
⑩ 根據大腦持有的這三個機能，由兩眼所映照出來的影像會**變成一個**。⑪ 所以，我們能夠看得到
立體的事物。

Why can you see clearly with glasses?

① The human eye is like a convex lens. ② The focal distance changes according to the thickness of the lens. ③ You usually move a magnifying glass back and forth when you look at something small through it to adjust the distance between the thing you want to see and your eyes equal to the focal distance.

④ You can not bring your eye into focus by adjusting the distance between the crystal lens and the retina, which receives images made by the light passing through the crystal lens, because the distance between the lens and the retina can not be changed in the eye. ⑤ Therefore you adjust the focal distance by changing the thickness of the crystal lens by the power of a muscle in the ciliary body.

⑥ A near sighted person can not focus on a distant thing because he or she can not make the crystal lens thin. ⑦ When the person sees through the glasses of a concave lens, he or she can see a distant thing clearly because this is in the same state as it is when he or she makes the crystal lens, the convex lens, thin.

⑧ On the other hand, a far sighted person wears glasses with a convex lens because the lens of his or her eye is the opposite of that of a near sighted person. (see p.145)

問54　メガネをかけると、なぜよく見えるようになるの？

答 ①**人間の眼は凸レンズのようなものです。**②レンズは厚さによって**焦点距離**も変わります。③**虫眼鏡**で**小さいものを見るとき**レンズの**位置**を調節するのは、見る対象と眼の位置をレンズの焦点距離に合わせるためです。

④人間の眼は、レンズである**水晶体**とレンズを**通過した**光が像を結ぶ**網膜**が**一体化し**ているため、それらの距離を**調整してピントを合わせる**ことができません。⑤そこで、**毛様体**の中にある**筋肉の力**で水晶体の厚さを変えて、焦点距離を調整します。
⑥**近視の人**は、水晶体を薄くできないため、**遠くに焦点を合わせられません。**⑦**凹レン**ズのメガネをかけると、水晶体の凸レンズを**薄くしたのと同じ状態**になり、遠くもはっきり見えるようになります。

⑧一方、**遠視の人**は近視の人と**逆の状態**なので、凸レンズのメガネをかけます。

問54　為什麼戴眼鏡才能看得清楚呢？

答 ①人類的眼睛就像是一個凸透鏡。②鏡片根據不同厚度，**焦點距離**也會改變。③當用**放大鏡**觀察微小物品時，會移動鏡片的位置，目的是在觀察對象與眼睛的**位置**間調整適當的焦點距離。④人類的眼睛是由身為鏡頭的**水晶體**，與光線**通過**鏡頭後形成**影像的視網膜**為一整體，所以**無法調整**它們的距離來對焦。
⑤**然而**，卻可以利用位於**睫狀體**中的**肌肉力量**，來改變水晶體的厚度、調整焦點距離。⑥**近視的人**由於水晶體無法變薄，**無法在遠處對焦**。⑦戴上**凹透鏡**的眼鏡後，會形成像水晶體的凸透鏡變薄一樣的**相同狀態**，以至於遠處也能**清楚地**看見。
⑧另一方面，因為**遠視的人**與近視的人**狀態相反**，所以戴上的是凸透鏡的眼鏡。

Why don't we see things double though we have two eyes?

眼は二つあるのに、なぜものは一つに見えるの？
為什麼我們有兩隻眼睛，卻只看得見一個物體呢？

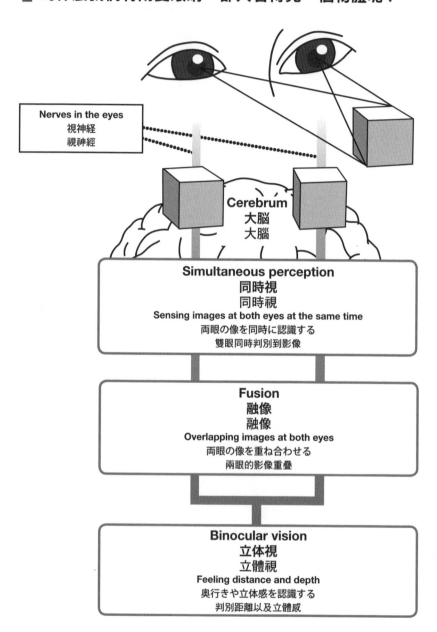

Nerves in the eyes
視神経
視神經

Cerebrum
大脳
大腦

Simultaneous perception
同時視
同時視
Sensing images at both eyes at the same time
両眼の像を同時に認識する
雙眼同時判別到影像

Fusion
融像
融像
Overlapping images at both eyes
両眼の像を重ね合わせる
兩眼的影像重疊

Binocular vision
立体視
立體視
Feeling distance and depth
奥行きや立体感を認識する
判別距離以及立體感

Glasses correct the focus of the crystal lens.

メガネは水晶体の焦点を補正する
眼鏡可以校正水晶體的焦點

Good eye
正常な眼
正常的眼球

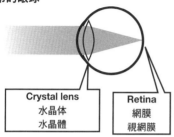

Crystal lens
水晶体
水晶體

Retina
網膜
視網膜

Light makes images on the retina clearly because crystal lens, the lens in the eye, has a focal point on the retina.

眼のレンズである水晶体の焦点が網膜上にあるので、光は網膜にはっきりと像を結ぶ

由於身為眼睛鏡頭的水晶體，在視網膜上形成焦點，光線在視網膜上結成清楚的影像。

Near sight
近視
近視

They cannot make the crystal lens thin so the focal point is before the retina.

水晶体を薄くできないので、焦点が網膜よりも手前になる

水晶體無法變薄，導致焦點落在視網膜稍前方

They can make the focal point far with a concave lens and see clearly.

凹レンズで焦点を後ろにずらすと、はっきりと見えるようになる

運用凹透鏡將焦點往後移，就能看得清楚。

Far sight
遠視
遠視

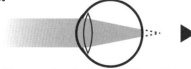 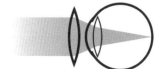

They cannot make the crystal lens thick so the focal point is after the retina.

水晶体を厚くできないので、焦点が網膜よりも奥になる

水晶體無法變厚，導致焦點落在視網膜稍後方

They can make the focal point near with a convex lens and see clearly.

凸レンズで焦点を手前にずらすと、はっきりと見えるようになる

運用凸透鏡將焦點往前移，就能看得清楚。

Q 55　How do they check the blood types, A, B, O, and AB?

①Human blood is classified into four types, A, B, O, and
　　　人類的　　血液　　被分類為…

AB. ②They are determined by the red blood cells and
　　　　　　　　　　　　　　　　　　　　紅血球

protein in blood serum. ③When you mix two different types of
蛋白質　　血清　　　　　　　　　　　　　混合　　　　　　　　類型

blood, you can find that some pairs clot and others do not clot.
　　　　　　　　　　　　　　　　　凝固

④Red blood cells in blood have A antigens or B antigens in them.
　　　　　　　　　　　　　　　　A抗原

⑤A person with A type blood has A antigen and a person with B
A型的人

type blood has a B antigen. ⑥A person with O type blood has

neither of them and a person who has both of them is one with AB
沒有其中的任何一個

type blood.

⑦Blood serum, the liquid portion of blood, has two kinds of
　　　　　　　　　　液體成分

antibodies, α and β, which make red blood cells together in
抗體　　　　　　　　　　　　　　　　　讓…凝固

reaction to the A antigen or the B antigen. ⑧A person with A type
與…產生反應

blood has β and a person with A type blood has α. ⑨A person

with O type blood has both of them. ⑩A person with AB type blood
　　　　　　　　　　　雙方兩者都有

has neither of them.

⑪When you mix two different types of blood, some pairs clot and

others do not clot and you can determine the blood type.

問 55　どうやって、Ａ、Ｂ、Ｏ、ＡＢの血液型を調べるの？

答　①人はＡ、Ｂ、Ｏ、ＡＢの4つの**血液型**のどれかに**分類されます**。②それは**赤血球**
　　　　　　　　　　　　　　　血液類型　　　　　　　　　　　　　　　　　　　　　　　　紅血球

と**血清**の中の**タンパク質**で決まっています。③**血液**を混ぜ合わせてみると、固まる組
　　血清　　　　　　　　　　　　　　　　　　　　　　血液

み合わせと固まらない組み合わせがあることがわかります。

④血液の中の**赤血球**には、**Ａ抗原**またはＢ抗原が含まれています。⑤**Ａ型の人**はＡ抗
　　　　　　　紅血球　　　　抗原

原、Ｂ型の人はＢ抗原を持っています。⑥**Ｏ型の人はどちらの抗原も持っていない**人

で、両方の抗原を持っている人はＡＢ型です。
　　雙方

⑦血液の**液体成分**である**血清**にはＡ、Ｂそれぞれの抗原に**反応して赤血球を凝集させ**
　　　　　液體成分　　　　　血清　　　　　　　　　　　　　　　　反應　　　　　　　凝固

る物質である、二種類の**抗体** α と β が含まれています。⑧Ａ型の人はβを、Ｂ型の人
　　　　　　　　　　　　　　抗體 アルファ ベータ

はαを持っています。⑨Ｏ型の人は**両方**の抗体を持っています。⑩ＡＢ型の人はどちら
　　　　　　　　　　　　　　　　　　両種

の抗体も持っていません。

⑪混ぜ合わせたときの凝集の様子から血液型を特定するのです。
　　　　　　　　　　　凝固　　状態　　　　　　　　特定

問 55　要如何判別Ａ、Ｂ、Ｏ、ＡＢ四種血型？

答　①**人類的血液被分類**為Ａ、Ｂ、Ｏ、ＡＢ四種類型。②那是由紅血球與血清中的蛋白質所決定
的。③試著將血液**混合**，就會發現有一部分會**凝固**、另一部分則不會凝固。
④血液之中的紅血球裡，含有**Ａ抗原**或是Ｂ抗原。⑤**Ａ型的人**擁有Ａ抗原、Ｂ型的人擁有Ｂ抗原。
⑥Ｏ型的人**沒有**任何抗原、擁有雙方抗原的人則是ＡＢ型。
⑦血液中的**液體成分**血清，含有 α 、 β 兩種**抗體**，是會各自與Ａ、Ｂ抗原產生反應、使紅血球**凝固**
的物質。⑧**Ａ型的人**擁有**β**抗體，Ｂ型的人擁有 α 抗體。⑨Ｏ型的人擁有**兩種**抗體。⑩ＡＢ型的人則
沒有任何一種抗體。
⑪當你混合兩種類型的血液，由血液凝固的狀態可以判斷出特定的血型。

How mysterious human body is!

人の体はこんなに神秘的!
人類的身體是如此的奧秘！

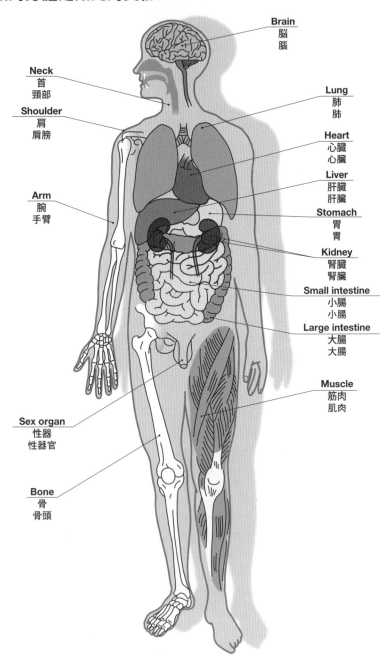

Brain
脳
腦

Neck
首
頸部

Shoulder
肩
肩膀

Arm
腕
手臂

Lung
肺
肺

Heart
心臓
心臟

Liver
肝臓
肝臟

Stomach
胃
胃

Kidney
腎臓
腎臟

Small intestine
小腸
小腸

Large intestine
大腸
大腸

Muscle
筋肉
肌肉

Sex organ
性器
性器官

Bone
骨
骨頭

Chapter 5

Questions
about
the things around us

第 5 章

身の周りの疑問
生活周遭的疑問

Q56 Why can microwave ovens heat up food?

①There is an antenna-like thing, called a magnetron, in
像線圈的裝置　　　　　　　　　　磁電管
the microwave ovens which gives off microwaves.
微波爐　　　　　　　　　釋放出　微波

②Any food has a lot of water molecules in it. ③A water molecule is
任何食物　　　　　　水分子
made up of two positively charged hydrogen atoms and one
由…所組成　　帶正電的　　　　氫原子
negatively charged oxygen atom. ④When a microwave hits the
帶負電的　　氧原子
water molecules, they vibrate and their temperature goes up
振動　　　他們的溫度
because of friction. ⑤That is why food will be heated. ⑥Microwaves
因為摩擦　　　因為這個原因　　　溫熱食物
of this kind of the oven pass through dishes, which have no water
穿過　　食器
in them, so dishes will not be heated. ⑦Therefore only food, which

has water in it, will be heated and become hot.

⑧The number of times an electric wave vibrates is called frequency.
回數　　　　　　　電波　　　　　　　　周波數
⑨The microwaves of this kind of oven have a frequency of 2,450
周波數為2450兆赫
megahertz, and positive charges and negative charges alternate
電荷　　　　　　　　　　　　使輪流…
with each other 2.45 billon times
24億5000萬次
per second. ⑩This frequency
每秒鐘
is suitable for vibrating water
適合…
molecules and heating food. (See
溫熱
p.162)

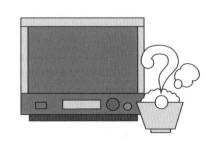

問56　電子レンジでどうして、料理が温まるの？

答 ①電子レンジは内部に「マグネトロン」というアンテナのようなものが設置され<ruby>電磁<rt></rt></ruby> <ruby>内部<rt></rt></ruby> <ruby>装置<rt></rt></ruby>
ていて、ここから**マイクロ波**という電波を**出します**。
<ruby>電磁波<rt></rt></ruby>

②**食べ物**にはたくさんの**水の分子**が含まれています。③水の分子は、**プラスの電気を帯**
<ruby>電<rt></rt></ruby>
びた**水素原子**が2個と**マイナスの電気を帯びた酸素原子**が1個からできています。④マ
<ruby>氫原子<rt></rt></ruby> <ruby>氧素原子<rt></rt></ruby>
イクロ波を水の分子に当てると、**揺さぶられて摩擦熱**が生じて温度が上がります。⑤
<ruby>摩擦生熱<rt></rt></ruby> <ruby>温度<rt></rt></ruby>
その結果、食べ物が**温められます**。⑥電子レンジのマイクロ波は水分を含まない**食器**
<ruby>食器<rt></rt></ruby>
を、**通り抜ける**ので温まりません。⑦そのため、水分を含む食べものだけ、発熱して
<ruby>加熱<rt></rt></ruby>
温まります。

⑧**電波が振動する回数**を「**周波数**」といいます。⑨電子レンジのマイクロ波の**周波数**は
<ruby>電波<rt></rt></ruby> <ruby>振動<rt></rt></ruby> <ruby>回數<rt></rt></ruby> <ruby>周波數<rt></rt></ruby>
2450メガヘルツで、**1秒間に24億5000万回**、プラスとマイナスが交互に**入れ替わり**
<ruby>交替<rt></rt></ruby>
ます。⑩この周波数が、水の分子を振動させ食品の内部まで**温める**のに**適している**の
<ruby>振動<rt></rt></ruby> <ruby>食物<rt></rt></ruby> <ruby>內部<rt></rt></ruby>
です。

問56　為什麼微波爐可以溫熱食物？

答 ①在**微波爐**的內部，有著稱之為「**磁電管**」的加熱線圈裝置，由這個線圈**發出**稱為**微波**的電磁波。
②**食物**裡含有大量的**水分子**。③水分子是由二個帶正電的氫原子與一個帶負電的氧原子所組合而成。④當微波碰到水分子時，**會產生振動並摩擦生熱、使溫度上升**。⑤因為這個原因，所以能夠**溫熱食物**。⑥微波爐的微波會**穿透**不含水分的**食器**，所以不會變熱。⑦因此，只有含有水分的食物才能加熱變熟。
⑧**電波振動的回數**，稱之為「**周波數**」。⑨微波爐的微波周波數為**2450兆赫**，每秒鐘可以正負極交替振動**24億5000萬次**。⑩這個周波數**剛好適合**振動水分子直到食物**變熟**。

Q57 Why do eggs get hard when they are boiled?

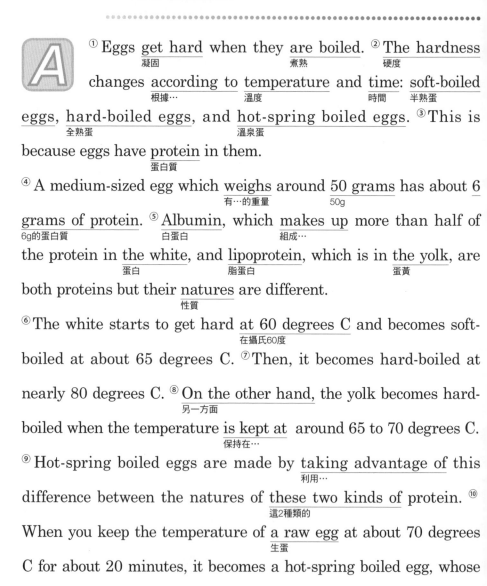

①Eggs get hard when they are boiled. ②The hardness changes according to temperature and time: soft-boiled eggs, hard-boiled eggs, and hot-spring boiled eggs. ③This is because eggs have protein in them.

④A medium-sized egg which weighs around 50 grams has about 6 grams of protein. ⑤Albumin, which makes up more than half of the protein in the white, and lipoprotein, which is in the yolk, are both proteins but their natures are different.

⑥The white starts to get hard at 60 degrees C and becomes soft-boiled at about 65 degrees C. ⑦Then, it becomes hard-boiled at nearly 80 degrees C. ⑧On the other hand, the yolk becomes hard-boiled when the temperature is kept at around 65 to 70 degrees C. ⑨Hot-spring boiled eggs are made by taking advantage of this difference between the natures of these two kinds of protein. ⑩When you keep the temperature of a raw egg at about 70 degrees C for about 20 minutes, it becomes a hot-spring boiled egg, whose white is softer than its yolk.

問57　卵はどうして、ゆでると固くなるの？

答 ①卵をゆでると、**固くなります**。②温度と時間を変えると「半熟卵」、「固ゆで卵」、

「温泉卵」と、固さも変化します。③これは、卵が**タンパク質**を含んでいるからです。

④卵に含まれるタンパク質は、中ぐらいのもの1個（約**50グラム**）につき、約**6グラム**程度で

す。⑤そして、卵白に含まれているタンパク質の半分以上を占める**アルブミン**と卵黄に含ま

れるリポタンパク質は、同じタンパク質でありながら異なる**性質**を持っています。

⑥卵白は約**60度**で固まり始め、およそ65度で半熟になります。⑦さらに、80度に近くになる

と固まります。⑧**一方**、黄身は約65度から70度くらいに**保つ**と、完全に固まります。

⑨**こうした2つのタンパク質の性質の違いを利用してつくる**のが、温泉卵です。⑩**生卵**をおよ

そ70度の温度で20分程度保温することにより、卵黄**よりも**卵白が**柔らかい**、温泉卵になる

のです。

問57　為什麼雞蛋煮熟之後會凝固？

答 ①雞蛋煮熟之後就會凝固。②隨著**溫度**與時間的變化，會形成「**半熟蛋**」、「**全熟蛋**」、
「**溫泉蛋**」等等不同的硬度。③這是因為雞蛋裡含有**蛋白質**的緣故。④雞蛋裡所包含的蛋
白質，每個中型蛋（50g）約含有6g左右。⑤然而，佔有蛋白所含有的蛋白質一半以上的**白蛋**
白，與蛋黃裡所含有的脂蛋白，雖然一樣是蛋白質，卻各有不同的**性質**。
⑥蛋白約在攝氏**60度**左右開始凝固，大概65度會變成半熟。⑦接下來，接近80度就會凝固。⑧**另**
一方面，蛋黃**保持**在65度到70度左右，就會完全凝固。⑨利用這2種蛋白質的性質差異，所製作
出來的就是溫泉蛋。⑩將**生蛋**以70度的溫度保溫20分鐘左右，蛋白**會比蛋黃柔軟**，做成溫泉蛋。

Q58 Why does plastic cling film stick to something?

① Plastic cling film has no glue on it. ② However, it sticks
　　保鮮膜　　　　　　　　　　　黏著劑

close to a plate or something like that.
緊緊地包覆在… 盤子

③ Why can it do so?

④ One of the reasons of this is that plastic cling film sticks close to

something hard such as a plate because it is very thin and soft.
堅硬的物品　　　　　　　　　　　　　　　　　　薄　　　柔軟

⑤ When a molecule of something gets close to another molecule,
　　　　　　分子　　　　　　　　　　接近…

they try to pull each other and stick to together. ⑥ This is called
　　　　　吸引　　　　　　　　　黏住

intermolecular force. ⑦ When you put a paper box on top of another
分子間作用力　　　　　　　　　　放在…之上

paper box, this force acts between two molecules close to each
　　　　　　　　　起作用　　　　　　　　　　互相接近

other.

⑧ Thin plastic cling film made from soft material sticks to the rim of
　　　　　　　　　　　　　　　　　素材　　　　　　　　邊緣

a plate or a cup. ⑨ It stick to the plate because of the intermolecular

force.

問58　ラップはなぜ、ぴったりくっつくの?

答 ①**ラップ**には**糊**(のり)はついていません。②にもかかわらず、お皿などに**ぴったりと**
黏著劑

密着させることができます。③これは、なぜでしょうか?
包覆

④その**理由**の一つとして、ラップが**薄くて柔らかく**、食器のような**硬いもの**(かた)にぴった
理由　　　　　　　　　　　　　　　　食器

りと密着することがあげられます。

⑤ものの**分子**と分子が**近づく**と、お互い**にくっつこう**とする力が働きます。
分子

⑥「**分子間力**」と呼ばれるものです。⑦紙の箱を二つ重ねたときにも、**接近した**分子の
分子間作用力　　　　　　　　　　　　　　　　　　　　　接近

間でこの力が**働き**ますが、**表面**に**凹凸**(おうとつ)があるため**非常**に弱いものです。
表面　凹凸　　　　　　非常

⑧柔らかい**素材**でできた薄いラップは、食器の**ふち**に密着します。⑨その部分に分子間
素材　　　　　　　　　　　　　　　　　　　　　　　　部位

力が働くため、食器にぴったりとくっつくのです。

問58　保鮮膜為什麼可以緊緊地包覆著?

 答 ①保鮮膜上並沒有沾黏著劑。②雖然如此,還是能夠緊緊地包覆在盤子等等的容器上。③這是為什麼呢?

④其中的一個理由是,保鮮膜既薄且柔軟,能夠緊緊地貼合在如食器這般堅硬的物品上。⑤當物品的分子與分子間相近時,彼此會相互吸引並相黏在一起。⑥這被稱之為「分子間作用力」。⑦兩個紙箱重疊的時候,兩個接近的分子當中雖然也有這個作用力在,但是因為表面的凹凸不平以至於變得薄弱。⑧由柔軟素材製作而成的薄薄的保鮮膜,能緊緊貼住食器的邊緣。⑨在這個部位由於分子間作用力的運作,所以能夠緊緊地包覆住食器。

Why do tears come in your eyes when you slice an onion?

① If you put an unsliced whole onion close to your eyes,
擺、放　　　還沒切的整顆洋蔥

tears do not come in your eyes.
流淚　　　　從眼睛流出

② However, when you slice an onion, it stings your eyes and you
切　　　　　刺激

shed tears. ③ Why does this happen?
流淚

④ Onions have an element, isoalliin, in them. ⑤ Isoalliin does not
成分　　　　異蒜氨酸

become volatile at room temperature and it does not sting your
揮發性　　　在常溫

eyes. ⑥ When you slice an onion, isoalliin changes into
改變

thiosulfinate because of an enzyme named alliinase. ⑦ In an
硫代亞磺酸酯　　　　　　酵素　　　　　　　蒜氨酸酶

unsliced onion, isoalliin and alliinase are in different cells, so they
在不同的細胞內

do not bind with each other.
結合

⑧ When you slice an onion, these cells are broken and these two

elements changes into thiosulfinate.

⑨ Thiosulfinate is irritating and is likely
刺激　　　　　容易

to become volatile. ⑩ This element comes
進入

into your eyes and you shed tears.

問59　タマネギを刻むと、どうして涙が出るの？

答 ①切っていない、**丸ごとのタマネギを眼に近づけても、涙が出る**ということはありません。②ところが、タマネギを**刻む**と眼にしみて、**涙が出てきます**。③これは、なぜでしょうか？

④タマネギには**イソアリイン**という**成分**が含まれています。⑤イソアリインは**揮発性**がなく、眼にしみることはありません。⑥タマネギを刻むと**アリナーゼ**という**酵素**の働きで、イソアリインが**チオスルフィネートに変化します**。⑦刻まない状態のタマネギでは、イソアリインとアリナーゼは**別の細胞にあり、混ざる**ことはありません。⑧刻むことによって細胞が壊れ、この二つが混じってチオスルフィネートができるのです。

⑨チオスルフィネートは**刺激性**があり、揮発**しやすい性質**があります。⑩これが眼に入ることによって、涙が出るのです。

問59　切洋蔥的時候為什麼會流眼淚？

答 ①還沒切好、整顆的洋蔥即使將它拿靠近眼睛，也不會有**流淚**的情形發生。②可是，一**切**洋蔥就會**刺激**到眼睛、**流出眼淚**。③這是為什麼呢？④洋蔥含有稱之為**異蒜氨酸**的成分。⑤異蒜氨酸沒有**揮發性**，並不會刺激眼睛。⑥在切洋蔥的時候，因為稱之為**蒜氨酸酶**的酵素運作，會將異蒜氨酸反應為**硫代亞磺酸酯**。⑦在還沒有切碎狀態的洋蔥裡，異蒜氨酸與蒜氨酸酶各自在**不同的細胞內，並不會混合**。⑧因為切碎之後會破壞細胞，這兩種物質混合之後就會產生硫代亞磺酸酯。⑨硫代亞磺酸酯具有**刺激性**與**容易揮發**的性質。⑩當它一**滲入**眼睛的時候，就會流出眼淚。

Q60　　Why does food go bad?

① Food goes bad because the organic matters, especially
腐敗　　　　　　　　　　　　　有機物質

protein, in it are broken down by bacteria called putrid
蛋白質　　　　　分解　　　　　　　　　細菌　　　　腐敗菌

bacteria and other microorganisms.
　　　　　　　　　　微生物

② There are a lot of bacteria in the air. ③ There are also a lot of

bacteria on your hands. ④ When these bacteria move to food, they

began to increase. ⑤ Food for human beings is so nutritious for
繁殖　　　　　　　　　　　　　　相當地　高營養價值

bacteria that they can increase easily in it.

⑥ When these bacteria break down protein in the food, it gives off
釋放…

smelly gas such as ammonia and hydrogen sulfide. ⑦ Some kind of
惡臭　　　　　　阿摩尼亞　　　硫化氫

bacteria make toxins which cause food poisoning. ⑧ This state is
毒素　　　　　引起…　食物中毒　　　　　狀態

what you say meat or fish goes bad.
這就是你所謂的…

⑨ Rice and bread, which have almost no protein in them,

sometimes go bad too. ⑩ This is caused mainly by bacteria called
主要是藉由…

grass bacilli. ⑪ The grass bacilli are related to
枯草桿菌　　　　　　　　　與…有親緣關係的

bacillus subtilis and they are everywhere.
納豆菌　　　　　　　　到處都有

⑫ Even when rice goes bad by these bacteria,
即使…當

the rice does not have a very bad smell and
氣味

these bacteria do not make many toxins.
幾乎不會

問60　　　　　　　　　　　食べ物はなぜ、腐るの？

答 ①「食べ物が**腐る**」のは、食べ物の中の**有機物**、特に**タンパク質**が**腐敗菌**と呼ば
（有機物質）　　　　　　　　　　　　　　　（腐敗菌）
れる**細菌**やその他の**微生物**によって**分解される**からです。
（細菌）　　　　（微生物）
②空気中には、細菌が漂っています。③また、人間の手にも多くの細菌がついていま
（飄浮）
す。④これらが食べ物につくと、**繁殖**を始めます。⑤人間の食べ物は、細菌に**とっても**
（繁殖）
栄養が豊かで繁殖しやすい**場所**だからです。⑥これらが食べ物の中のタンパク質を分
（營養）　　　　　　　（場所）
解すると、**アンモニア**や**硫化水素**など、**悪臭**の原因になるガス**を出します**。⑦菌の
（硫化氫）　　　（惡臭）
種類によっては、**食中毒**の原因となる**毒素**を出すこともあります。⑧これが、肉や魚
（種類）　　　　　　（食物中毒）
などが腐った**状態**です。
（狀態）
⑨タンパク質をほとんど含まないご飯やパンが腐ることもあります。⑩その原因となる

のは、**主に枯草菌**という細菌です。⑪**納豆菌の仲間である**枯草菌は、ごくありふれた
（枯草桿菌）　　（細菌）　　　（納豆菌）（相似）
細菌です。⑫この菌でご飯が腐敗しても、極端な悪臭や毒素を出すことは**あまりあり**
（腐敗）　　　（極端）　　　（毒素）
ません。

問60　　　　　　　　　　　食物為什麼會腐敗？

答 ①「食物腐敗」指的是食物中的**有機物質**，特別是被稱之為**蛋白質腐敗菌**的**細菌**以及其他
的**微生物分解**所導致。
②在空氣中飄浮著許多細菌。③此外，人類的手上也附著了許多細菌。④這些細菌接觸到食物
後，會開始**繁殖**。⑤而人類的食物對於細菌來說，是**營養豐富**且容易繁殖的場所。⑥當這些細菌
將食物中的蛋白質分解時，會釋放**阿摩尼亞**或**硫化氫**等等，**造成惡臭原因的瓦斯出來**。⑦依照細
菌種類的不同，也會釋放出會造成**食物中毒**的**毒素**。⑧這就是肉或魚等等腐敗的**狀態**。
⑨幾乎不含蛋白質的飯或麵包也會腐敗。⑩造成這個原因，**主要是稱之為枯草桿菌**的細菌。⑪與
納豆菌相似的枯草桿菌，到處都存在著。⑫這個細菌雖然會讓米飯腐敗，**卻幾乎不會產生極端的**
惡臭以及毒素。

Q61 Why is a drainpipe s-shaped?

① The drainpipes under the washstand, the kitchen sink, and so on are S-shaped so that water is always standing at their curve. ② It is true that there are various shapes but all the drainpipes are designed to keep some water in the middle of them in this way. ③ The water kept there is called seal water and it seals the drainpipe to keep away smells and bugs. ④ This system is called a trap.

⑤ Even if there is a trap in a drainpipe, it does not work when you let flow a large amount of water at one time and there is no seal water. ⑥ This is a phenomenon called self-siphon. ⑦ Siphon is a system that can fill a curved pipe with liquid and lead the liquid to the end point at a lower level than the starting point by way of a point at a higher level than the starting point.

⑧ When you let flow a large amount of water at one time, the drainpipe and trap work as a siphon and the seal water kept there flows

away. (see p.163)

問61　排水管の形はなぜ、Ｓ字型なの？

答①たとえば、**洗面所**や**台所**などの下の**配水管**はＳ字に曲がっていて、常に水がた

まるようになっています。②形は**さまざま**ですが、排水設備には必ずこのように、途

中に水をためておく仕掛けがあります。③たまっている水は「**封水**」と呼ばれ、排水

経路をふさいで臭いや虫などの**進入**を**防ぎ**ます。④この**仕掛け**は、**トラップ**と呼ばれ

ています。

⑤せっかくトラップがついていても、**大量の水**を**一気に流す**と水がなくなり、効かな

くなることがあります。⑥これは「**自己サイフォン**」と呼ばれる**現象**です。

⑦サイフォンとは、**曲がった管**の中で**液体**が途切れないようにして、ある地点から高

い地点を越え、**出発点より低い終点**まで液体を**導く**ようなしくみです。⑧大量の水を

流すと、配水管とトラップがサイフォンの状態になり、たまった封水が**流れ出て**しま

うのです。

問61　排水管的形狀為什麼是S型的呢？

答①舉例來說，**洗臉檯**或**廚房**等等，底下的**排水管**彎曲為S型，讓水經常保持著蓄積的狀態。
②雖然形狀有**各式各樣**，但是排水設備必須要有像這樣，在途中讓水蓄積的設計。③蓄積
起來的水稱之為「**水封**」，可以封住排水通路，**防止惡臭或害蟲侵入**。④這個設計稱之為**防臭
瓣**。
⑤雖然特意地設計出防臭瓣，一旦一口氣排入大量的水時，水封現象就會失效。⑥這個**現象**稱之為
「**自我虹吸**」。
⑦所謂的虹吸現象指的是，當由某個地點跨越到較高的地點時，將**彎曲水管**中的**液體**分段，把液
體**導引成終點比出發點還低**的狀態。⑧當流入大量的水時，排水管與防臭瓣會形成虹吸狀態，蓄
積的水封就會流出。

The microwave oven vibrates and heats water molecules in food with a microwave.

電子レンジはマイクロ波で、食べ物の水分子を振動させて温める
微波爐以微波振動食物中的水分子，使其溫熱

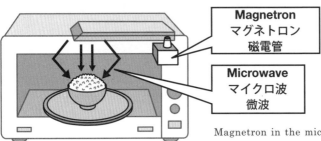

Magnetron
マグネトロン
磁電管

Microwave
マイクロ波
微波

Magnetron in the microwave oven gives off a microwave.
電子レンジでは、マグネトロンからマイクロ波を出している
微波爐由磁電管發射微波

Microwave
マイクロ波
微波

> **Microwave vibrates water molecules in food.**
> マイクロ波が食べ物の水分子を振動させる
> 微波會振動食物中的水分子

> **Heat is caused by the friction of water molecules and food is heated.**
> 水分子の振動で摩擦熱が生じ、食べ物が温まる
> 因水分子的振動而摩擦生熱，讓食物變熱

Why you mustn't let flow a large amount of water in the drainpipe at one time

排水口に大量の水を流してはいけない理由
不可以往排水口一次倒入大量的水的理由

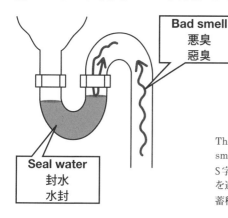

Bad smell
悪臭
惡臭

Seal water
封水
水封

The water kept at S-shaped pipe keeps away bad smell from the sewer.

S字部分にたまった水が、下水から流れ込んでくる悪臭を遮断している

蓄積在S型水管裡的水，可以阻隔由下水道傳來的惡臭。

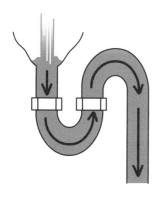

When you let flow a large amount of water at one time ...
大量の水を一気に流すと……
當大量的水一口氣倒入

The water kept there all flows away.
たまった水が、途切れることなく押し出される
蓄積的水無法阻隔而全部排出

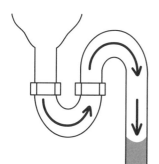

Water kept at S-shaped pipe also flows away by siphon effect.
サイフォンの効果で、S字部分にたまっていた封水まで、流れ出てしまう
由於虹吸效應，蓄積在S型水管裡的水也會被吸引而全部排出

Q62 What is the real color of shrimps?

①The striped pattern of shrimps in a fish shop is
條紋花樣　　　　　　　　　蝦子

blackish. ②However, the striped pattern of boiled
呈黑色的　　　　　　　　　　　　　　煮熟的

shrimps on the table is reddish. ③Why do they differ?
呈紅色的　　　　　　　不同

④Boiled shrimps looks red because they have a red coloring
看起來　　　　　　　　　　　　　　色素

matter, carotin, in them. ⑤When you hear of carotin, you may
胡蘿蔔素　　　　　　　　聽到

think of carrots or tomatoes. ⑥The carotin of shrimps and crabs
紅蘿蔔或番茄　　　　　　　　　　　　　　螃蟹

have is similar to that of carrots and tomatoes.
與…相似

⑦The color of the striped pattern of uncooked shrimps is also that
生的

of carotin but it looks blackish, not reddish.

⑧That is because carotin binds with protein here. ⑨The carotin
與…結合　　蛋白質

bound with protein looks dark red or green.
與…結合後

⑩When you boil shrimps, the binding between carotin and protein
結合

comes undone. ⑪As a result, carotin
脫落　　　　　結果

is back in its original color and shrimps
回復成原來的顏色

come to look red.

問62　エビの色は、本当は何色なの？

答 ①魚屋の店頭に並んでいる**エビの縞模様**は、黒っぽい色をしています。②ところ

が、食卓に並ぶ、**ゆでたエビの縞模様は赤っぽい色**をしています。③これは一体、ど

ういうことなのでしょうか？

④ゆでたエビやカニが赤く**見える**のは、**カロチン**という赤い色素を含んでいるからで

す。⑤カロチンといえば、**ニンジンやトマト**が思い浮かびます。⑥エビや**カニ**のカロチ

ンは、ニンジンやトマトに含まれるカロチンに**近い**ものです。

⑦**生のエビの縞模様もカロチンですが、赤ではなく黒っぽく見えます。⑧なぜなら、カ

ロチンが**タンパク質と結合している**からです。⑨**タンパク質と結合した**カロチンは、

暗い赤や緑色になります。

⑩エビをゆでると、熱によってカロチンとタンパク質の**結合**が**はずれます**。⑪その

結果、カロチンは**本来の色に変化し、エビが赤く見えるようになるのです。

問62　蝦子原本應該是什麼顏色呢？

答 ①排列在魚舖店頭的**蝦子**，呈現著**黑色條紋的花樣**。②然而，排列在餐桌上、**煮熟後的蝦子卻呈現著鮮紅色條紋的花樣**。③這究竟是為什麼不一樣呢？
④煮熟後的蝦子或螃蟹會呈現鮮紅色，是因為牠們含有稱為**胡蘿蔔素**的色素的緣故。⑤一說到胡蘿蔔素，就會想起**紅蘿蔔或是番茄**。⑥蝦子或**螃蟹**含有的紅蘿蔔素跟紅蘿蔔與番茄所含有的紅蘿蔔素是**相似的**物質。
⑦**生的蝦子的條紋花樣雖然也有紅蘿蔔素，但看起來不是鮮紅色而呈現黑色。⑧那是因為紅蘿蔔素與**蛋白質結合在一起**的緣故。⑨與蛋白質**結合後**的紅蘿蔔素會呈現深紅或是綠色。
⑩蝦子一煮熟，因為加熱破壞了紅蘿蔔素與蛋白質的**結合**。⑪結果，紅蘿蔔素**變化成原來的顏色**，蝦子看起來就成了鮮紅色。

Q63 What is the difference between fish with white meat and fish with red meat?

① Human muscle fibers are made up of red muscle and
人類的　　肌肉纖維　　　　　由…組成　　　　紅色肌肉

white muscle. ② Red muscle, which has a lot of red
白色肌肉

matter named myoglobin to store oxygen, has stamina. ③ White
紅色物質　　　　肌紅蛋白　　儲存　氧氣　　持久力

muscle, which has instantaneous power and can contract quickly,
爆發力　　　　　　　　收縮

has little staying power and gets tired soon. ④ Human muscles are

a mixture of red muscle and white muscle, so they look pink.
混合

⑤ The difference between fish with white meat and fish with red
…與…兩者間的不同　　　　白身魚　　　　　　赤身魚

meat is a similar story. ⑥ Tuna, mackerel, and sardine need to
與…相似　　　　　鮪魚　　青花魚　　　沙丁魚　　必須要…

have stamina because they swim long distances. ⑦ That is why
長距離

they have a lot of myoglobin in their muscles.

⑧ Flounder, sea bream, and so on, which don't swim long distances,
比目魚　　鯛魚

need to get away quickly when they are attacked. ⑨ That is why
逃跑　　　　　　　　　　被攻擊

their muscles are white and have instantaneous power.

⑩ Horse mackerel are sometimes classified with fish with white
竹筴魚　　　　　　　　　被歸類為…

meat because most of their muscles are white, but they actually
實際上

swim long distances and they are one of the types fish with red

meat.

問63　白身の魚と赤身の魚は、どう違うの？

答 ①**人間の筋肉繊維**には、赤い色をした**赤筋**と白い色をした**白筋**があります。②赤筋は
　　　人類　　肌肉纖維　　　　　　　　　紅色肌肉　　　　　　　白色肌肉

ミオグロビンという、**酸素を蓄える**働きがある**赤い物質**をたくさん含んでいて、**持久力**が
　　　　　　　　　　　氧氣　儲存　　　　　　　　　　　　　　　持久力

あります。③白筋は**瞬発力**があって、すばやく**収縮します**が、持久力に乏しく、すぐに疲れて
　　　　　　瞬間爆發力　　　　　　　収縮　　　　　　　　　　　乏

しまいます。④人間の筋肉は、この白筋と赤筋が**混じっている**のでピンク色をしています。

⑤**白身の魚**と**赤身の魚**の違いも、これと似ています。⑥**マグロやサバ、イワシ**のように**長距**
　　白身　　　　赤身　　　　　　　　　　　　　　　　　　　　　　　　　長距離

離を回遊する魚は、持久力**を必要とします**。⑦そのため、筋肉にミオグロビンが多く赤身で
　　迴游　　　　　　　　　　　必須要有

す。

⑧**ヒラメや鯛**のように回遊しない魚は、敵に**襲われた**ときにとっさに**逃げる**必要がありま
　　　　　鯛魚

す。⑨そのため、筋肉は瞬発力がある白身です。
　　　　　　　　　　瞬間爆發力

⑩**アジ**は、白身が多いため白身魚に**分類される**ことがありますが、**実際には**回遊魚であり、
　　　　　　　　　　　　　　　　歸類　　　　　　　　　　實際

赤身魚の仲間です。
　　　一類

問63　白身魚與赤身魚有什麼不一樣？

答 ①**人類的肌肉纖維裡**，有呈現紅色的**紅色肌肉**與呈現白色的**白色肌肉**。②紅色肌肉裡含有
　　大量被稱為**肌紅蛋白**，有儲存氧氣功能的**物質**，具有**持久力**。③白色肌肉有**瞬間爆發力**，
雖然能夠**瞬間收縮**、但缺乏持久力，容易立刻疲勞。④人類的肌肉因為由白色肌肉與紅色肌肉**組
合而成**，因此呈現粉紅色。⑤**白身魚與赤身魚之間的差異**，則與這個原因相似。⑥像**鮪魚、青花
魚、沙丁魚**這些**長距離**的迴游魚類，**必須要有**持久力。⑦因此肌肉裡含有大量的肌紅蛋白，呈現
紅色。⑧像**比目魚、鯛魚**這類非迴游魚類，當受到敵人**攻擊**時必須要瞬間**逃離**。⑨因此肌肉是呈
現擁有爆發力的白色。

⑩**竹筴魚**由於身上的肌肉多呈現白色，**被歸類為白身魚的一種**，**實際上**是屬於迴游魚類，歸類為
赤身魚的一類。

Q 64 Why do things burn when they are set on fire?

① Burning means that something binds with oxygen
燃燒　　　指的是…　　　　　　　與…結合　　氧氣

giving off light and heat in an intense manner. ② There
釋放　　　　　　　　　　　以強烈的方式

are a lot of combustible things around us such as paper, wooden
易燃物　　　　　　　　　　　　　　　　木製品

products and cloth. ③ However, they do not burn by themselves if
自己

you do not set fire to them or heat them. ④ Why don't they burn?
點火　　　　　　　加熱

⑤ Imagine that you move a lighter close to a dry wooden stick. ⑥ At
想像這個…　　　　　移動　打火機　　　　　乾燥的 木棒

first, the surface of the wooden stick becomes dark brown or black,
表面　　　　　　　　　　　　茶褐色

and then it gradually begins to burn with a flame.
漸漸地　　　　　　　　起火燃燒

⑦ This wooden stick does not burn simply by moving a fire close to
不僅僅藉由…

it because cool wood does not have any combustible element in it.
冷卻的　　　　　　　　　　易燃成分

⑧ If we heat it a lighter, the element of wood breaks down to
加熱　　　　　　　　　　　　　　　分解

become combustible gases such as carbon monoxide. ⑨ Then, these
可燃性的瓦斯　　　　　一氧化碳

gases start to burn with a flame.

⑩ Petroleum, gasoline, alcohol, and so on are easier to burn than
石油　　　汽油　　　酒精　　　　　　比…容易燃燒

wood or paper. ⑪ These materials release easily combustible gases
物質　　　釋放

even though they are not heated.
即使

問64　火をつけると、なぜ物が燃えるの？

答 ①**燃焼**とは、物質が**激しく**光や熱を出しながら酸素と結びつくことです。②私た
　　　燃燒　　　　物質　　　　　　　　　　　　　　氧氣

ちの身の周りには紙、**木製品**、布など、**燃えるもの**がたくさんあります。③にもかか
　　　　　　　　　木製品

わらず、これらは**火をつけたり**高温にしたりしないと燃えません。④これは、なぜで
　　　　　　　　　　　　　　加熱

しょうか？

⑤**ライターの火を乾いた木の棒に近づける**とします。⑥まず、木の**表面**がこげたような
　　　　　　　　　　かわ　　　　　　　　　　　　　　　　　　　表面

色になり、**しだいに炎を上げて燃え始めます。**

⑦火を近づけただけでは燃えない**理由**は、**冷えた**状態の木には、燃える**成分**がないか
　　　　　　　　　　　　　　　　理由　　冷卻　　狀態　　　　　　　成分

らです。⑧ライターの**火で熱する**ことにより、木の成分が**分解され**、一酸化炭素な
　　　　　　　　　　　　　　　　　　　　　　　　　分解　　　一酸化炭

ど、**可燃性のガス**になります。⑨ガスが燃え出すと、炎が上がります。
　　可燃性

⑩**石油、ガソリン、アルコール**などは、木や紙よりも**燃えやすい性質**があります。⑪こ
　　石油　　　　　　　　　　　　　　　　　　　　　　　　　性質

れらは熱せられなくても、**蒸発**するだけで**非常**に燃えやすいガスになるからです。
　　　　　　　　　　　　蒸發　　　　　　非常

問64　為什麼點火之後，物品會燃燒？

答 ①所謂的**燃燒**是指物質一面與**氧氣**結合，一面釋放出**強烈**的光和熱。②我們的生活周遭有
著大量的紙、**木製品**、布類等等**易燃物品**。③雖然如此，這些物品**不點火**或加熱，並不會
燃燒。④這是為什麼呢？
⑤試著將**打火機**的火接近乾燥的木棒。⑥首先，木頭的**表面**會呈現茶褐色，漸漸地燃起**火焰**並開
始燃燒。⑦不將火焰靠近便無法燃燒的理由是因為，**冷卻**狀態的木頭並沒有燃燒的成分。⑧經由
打火機的**火加熱**，木頭的成份**開始分解**，形成一氧化碳等等**可燃性的瓦斯**。⑨瓦斯釋放出來後便
點燃了火焰。
⑩**石油、汽油、酒精**等等含有比木頭及紙張**易燃**的性質。⑪這些物質即使不加熱也容易蒸發，形
成易燃的瓦斯。

Why don't pickled plums go bad though they are not dry?

① Food goes bad because of microorganisms such as
腐敗   微生物

bacteria. ② Almost all the microorganisms can not
細菌

increase without water. ③ Therefore people have made dry food by
增加   乾貨

putting it outside of the house and removing water in the sun
          在戶外   去除…

since early times in order to keep the food good. ④ However, why do
自古以來   為了…   保存

pickled plums go bad though they have water in them?
醃梅干   雖然

⑤ Water in food has two kinds: bond water and free water. ⑥ Bond
                                結合水   自由水

water binds with the elements of food chemically, so it does not
        結合   成分   化學的

evaporate by ordinary way of drying and it does not freeze until
蒸發   普通的乾燥法   凍結   直到接近

around minus 30 degrees C either. ⑦ On the other hand, free water

binds with the elements of food weakly, so it can move freely and
                                虛弱地   自由地移動

evaporate easily.

⑧ Microorganisms increase with this free water and they make

food go bad.

⑨ One of the reasons why pickled plums are not likely to go bad is
                                          不容易

that they have little free water in them other than that they are
                                        除了   強酸

highly acidic and they have salt. ⑩ Pickled plums sold in town
                          鹽分   市售的

sometimes go bad because they do not have so much salt.

問 65　梅干しは乾燥しているわけでもないのに、なぜ腐らないの？

答 ①食品が**腐る**のは、**細菌**などの**微生物**によるものです。②ほとんどの微生物は、<u>水分</u>
　　 食物　　　　　　　 細菌　　　　　 微生物　　　　　　　　　　　　　　　 水分

がないと**増える**ことができません。③そこで、**昔から食品を保存するために**、<u>天日に干して</u>
　　　　　　　　　　　　　　　　　　　　 食物　 保存　　　　　　　 てん び　　は
　　　　　　　　　　　　　　　　　　　　　　　　　　　　　　　　　 豔陽

水分**を除く乾物**がつくられてきました。④ところが、**梅干し**には水分が含まれているにもか
　　 のぞ　　　乾貨　　　　　　　　　　　　　　 醃梅干

かわらず、なぜ腐らないのでしょうか。

⑤食品に含まれる水分には、**結合水**と**自由水**の2種類があります。⑥結合水は、食品の**成分**
　　　　　　　　　　　　 結合水　　 自由水　　　　　　　　　　　　　　　　　　　　　 成分

と**化学的に結合**しており、**通常の乾燥法では蒸発せず**、**マイナス30度近く**にならないと**凍**
　 化學　　　　　　　　　　　　 乾燥法　 蒸發　　　　　　　　　　　　　　　　　　　　　 凍結

結もしません。⑦自由水は、食品との結合力が**弱く**、**自由に動く**ことができる水分で、**簡単**
　　　　　　　　　　　　　　　　　　　　　　　　 自由　　　　　　　　　　　　　　　 簡單

に蒸発します。⑧微生物はこの自由水によって増殖し、食品を腐敗させます。
　　　　　　　　　　　　　　　　　　　　　 繁殖　　　　　 腐敗

⑨**梅干し**が乾燥していないのに腐り**にくい**理由には、**強い酸と塩分**のほか、この自由水が少
　　　　　　 乾燥　　　　　　　　　　　　 理由　　　　 強酸　鹽分

ないという理由もあります。⑩**市販**の梅干しが腐る場合がありますが、塩分濃度が低く抑え
　　　　　　　　　　　　　　 市售　　　　　　　　　　　　　　　　 鹽分濃度

られているためです。

問 65　為什麼醃梅干即使不乾燥也不會腐敗？

答 ①食物會**腐敗**，是因為**細菌**等等的**微生物**所致。②幾乎所有的微生物，在沒有水分的情況下就無法**繁殖**。③因此，**自古以來為了保存食物**，會在豔陽下曝曬食物，**除去水分作成乾貨**。④但是，雖然**醃梅干**含有水分，為什麼卻不會腐敗呢？⑤食物裡面含有的水分，分成**結合水**與**自由水**2個種類。⑥結合水與食品的**成分**作化學結合，用普通的**乾燥法無法蒸發**，必須在負30度左右才**會凍結**。⑦自由水則與食品的結合力**弱**，屬於能夠**自由移動**的水分，簡單就能蒸發。⑧微生物就是經由自由水繁殖，讓食物腐敗。⑨醃梅干雖然不需要乾燥，卻**不容易腐敗的理由**，是因為除了**強酸**與**鹽分**之外，它幾乎不含這個自由水的緣故。⑩**市售**的醃梅干雖然也有腐敗的時候，那是因為它們缺乏足夠的鹽分。

Q66 Why water changes to ice when it is cooled?

① Not only in water but also in any materials, molecules have a power to pull one another. ② On the other hand, they have a power to push each other and try to keep a certain distance when they move too close.

③ This is a power called intermolecular force. ④ Other than this, molecules have a nature to move actively when they got energy.

⑤ The state of materials changes from gas to liquid, and then, from liquid to solid according to the balance between the molecular movement and the intermolecular force.

⑥ Gas is a state in which molecules have a large amount of energy and each of them move actively and freely. ⑦ Liquid is a state in which molecules pull one another by the intermolecular force and don't move so actively because the molecular movement goes down. ⑧ Solid is a state in which molecules stick close together and move little.

⑨ To cool water is to take heat energy away from water. ⑩ The molecular movement of water which is taken its energy away goes down. ⑪ Therefore the distance between molecules becomes small and the molecules stick close together to become solid by the intermolecular force. ⑫ This state is ice.

問66　水を冷やすと、なぜ氷になるの？

答 ①水に限らず、物質を構成する**分子**にはお互いに引きつけ合う力が働きます。②
近づきすぎると、今度は反発して遠ざけようとする力が働き、**一定の間隔を維持**しよ
うとします。③「**分子間力**」と呼ばれる力です。④**これとは別に、分子はエネルギーを**
得ると運動が激しくなる、という**性質**を持っています。
⑤**分子の運動**と分子間力の**バランス**によって、物質は**気体**から**液体**、液体から**固体**と
形を変えていきます。⑥気体は、分子が**大きな**エネルギーを持ち、激しく運動してバ
ラバラになっている状態です。⑦液体は分子運動が**減少**し、分子間力で互いに引き合
い、激しく運動しない状態です。⑧そして、固体は分子間力で分子が**しっかりくっつ**
き、大きな**運動をしない**状態です。

⑨「**水を冷やす**」とは、「**水から熱エネルギーを奪う**」ことです。⑩エネルギーを奪わ
れた水の分子運動は減少します。⑪そのため、分子間の距離も小さくなり、分子間力
によって、分子がしっかりとくっついて固体になります。⑫これが氷です。

問66　為什麼將水冷卻之後會結成冰？

答 ①不只是水，任何物質都是由**分子**相互牽引的力量而構成。②當過於靠近的時候，一個相
反的作用力會運作，讓它們**保持一定的距離**。③這種力量被稱之為「**分子間作用力**」。④
除此之外，分子還具有得到能量之後會激烈運動的**性質**。⑤根據**分子的運動**與分子間作用力的平
衡，物質的形態會有由**氣體**變成**液體**、液體變成**固體**的變化。⑥氣體的分子擁有**大量**的能量，激
烈的運動呈現自由的狀態。⑦液體的分子運動**減少**，以分子間作用力相互結合，處於非激烈運動
的狀態。⑧然後，固體因分子間作用力**緊緊地貼在一起**，處於**幾乎不運動**的狀態。⑨所謂的「將
水冷卻」，就是「**從水中奪取熱的能量**」。⑩能量被奪走之後，水的分子運動減少。⑪因此分子
間的距離也變小，因為分子間作用力，分子便緊緊地結合而形成了固體。⑫這就是冰塊。

Q 67 Why does it get cold in the refrigerator?

① As noted before, the state of materials changes from
如同前面所說　　　　　　　狀態　　　　　物質

solid to liquid, and then, from liquid to gas according to
固體　　液體　　　　　　　　　　　　　氣體　根據…

its energy and the intermolecular force.
　　　　　　　分子間作用力

② Pressure is also related to these changes. ③ When you put pressure
壓力　　　　　與…有關　　　　　　　　　　　　　　施加壓力在…上

on gas, the intermolecular force is likely to work because molecules
　　　　　　　　　　　　　　容易…　　　　　　　　分子

come close one another. ④ As a result, gas changes to liquid giving
靠近　　　　　　　　　　　　　　　　　　　　　　　釋放…

off energy, or heat.
　　　　　　熱能

⑤ When you ease pressure on it, the reverse takes place. ⑥ That
　　　　減輕　　　　　　　　相反的　　引起　　　　　總之

means molecules get cold and changes to gas taking in energy, or
　　　　　　　　冷卻　　　　　　　　　　　　拿取

heat.

⑦ These two theories are applied to the household refrigerator. ⑧ A
　　　　理論　適用…　　　　　　家用的　電冰箱

material called cooling medium is kept in the cooling unit of
　　　　　　冷媒　　　　　　保存在…　冷卻裝置

refrigerator. ⑨ When pressure is put on the cooling medium by

a pump that runs on a motor, the state of the cooling medium
幫浦　　　起動馬達

changes from gas to liquid giving off heat. ⑩ The liquid cooling

medium is shot up to the place of low pressure, so it changes to
　　　被噴出　　　　　　　　低壓

gas and evaporates. ⑪ It gets cold in the refrigerator because heat
　　　蒸發

is taken away in the process.
被…帶走　　在這個過程中

67 # 冷蔵庫はどうして、冷えるの？

答 ①**物質**は、それが持つエネルギーと**分子間力**によって固体、**液体**、**気体**と**状態**を
物質　　　　　　　　　　　　　　　　　分子間作用力　　　　　　　　　　　　　　形態

変えることは、すでに説明したとおりです。
　　　　　　　説明

②この変化には**圧力**も**関係します**。③**気体**に大きな**圧力**がかかると、**分子**が**接近**して分
　　　　　　　壓力　關係　　　　氣體　　　　　壓力　　　　　　　　　　接近

子間力が働き**やすく**なります。④その**結果**、**気体**はエネルギー（**熱**）を**出しながら**
　　　　　　　　　　　　　　　　　結果　　氣體

液体になります。
液體

⑤液体の圧力を下げると、**逆のこと**が**起こります**。⑥分子はエネルギー（熱）を使って

冷たくなりながら気体になります。

⑦**家庭用の電気冷蔵庫**は、これら二つの**原理**を**利用**したものです。⑧冷蔵庫の**冷却装置**
　　家用　　電冰箱　　　　　　　　　　　　　原理　利用　　　　　　　　　　　冷卻裝置

には、**冷媒**とよばれる物質が**閉じ込められて**います。⑨**モーターで動くポンプ**で冷媒
　　　冷媒　　　　　　　　　放入　　　　　　　　　　れいばい

に圧力をかけると、熱を出しながら気体から液体になります。⑩液体となった冷媒
　　　　　　　　　　　　　　　　　　　　　　　　液體

を、**圧力が低い**ところに**噴出すると**、**蒸発して**気体となります。⑪このとき熱が**奪わ**
　　壓力　　　　　　　噴出　　　蒸發

れるため、冷蔵庫は冷えるのです。

問67 # 為什麼冰箱可以冷卻？

答 ①**物質**會因為本身持有的能量與**分子間作用力**，改變為固體、**液體**或氣體的形態，這在前
篇已經有詳細說明。
②這個變化與**壓力**也有關係。③將較大的**壓力**施加在氣體上的時候，由於**分子接近**，分子間的作
用力**變得活躍**。④結果，氣體一面**釋放出能量（熱能）**，一面轉變成液體。⑤將液體的壓力降低
的時候，就會**引起相反的反應**。⑥分子一面使用能量（熱能）冷卻，一面轉變成氣體。
⑦**家用的電冰箱**，就是利用這兩個原理製造而成的。⑧冰箱的**冷卻裝置**裡，放入了稱之為**冷媒**的
物質。⑨當馬達起動幫浦對冷媒施加壓力，會一面釋放出熱能、一面由氣體轉變成液體。⑩將變
成液體的冷媒用**低壓**噴出時，就會**蒸發**變成氣體。⑪這個時候由於熱能被帶走了，冰箱就能夠降
溫。

When energy is taken away from water, its molecules stick close together and it changes into ice.

水のエネルギーを奪うと、
分子がしっかりくっついて氷になる
當水的能量被帶走
分子便緊緊地結合在一起形成冰

water vapor(gas)

水蒸気（気体）

水蒸氣（氣體）

The state in which molecules have a large amount of energy and each of them move actively and freely

分子が大きなエネルギーを持ち、激しく運動してバラバラになっている状態

氣體的分子擁有大量的能量，激烈地運動呈現自由的狀態

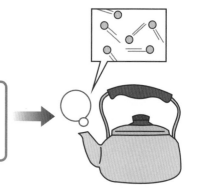

water(liquid)

水（液体）

水（液體）

The state in which molecules pull one another by the intermolecular force and do not move so actively because the molecular movement goes down

分子運動が減少し、分子間力で互いに引き合い、激しく運動しない状態

分子運動減少，以分子間作用力相互結合，處於非激烈運動的狀態

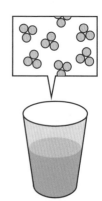

ice(solid)

氷（固体）

冰（固體）

The state in which molecules stick close together and move little

分子間力で分子がしっかりくっつき、大きな運動をしない状態

因分子間作用力緊緊地貼在一起，處於幾乎不運動的狀態

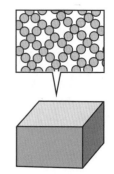

It gets cold in the refrigerator by the vaporization heat a liquid cooling medium gives off when it changes to gas.

電気冷蔵庫は、冷媒が気体になるときの気化熱で庫内を冷やす

冰箱利用冷媒轉變成氣體時的汽化熱，來冷卻冰庫。

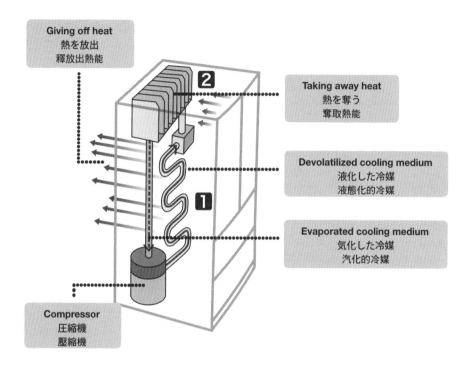

Giving off heat
熱を放出
釋放出熱能

Taking away heat
熱を奪う
奪取熱能

Devolatilized cooling medium
液化した冷媒
液態化的冷媒

Evaporated cooling medium
気化した冷媒
汽化的冷媒

Compressor
圧縮機
壓縮機

Gaseous cooling medium put on pressure by a compressor changes to liquid giving off heat.
気体状態の冷媒は、圧縮機で圧力をかけられ、熱を出しながら液体になる。
氣體狀態的冷媒是由壓縮機施加壓力，一面釋放熱能、一面轉變為液體。

2

Liquid cooling medium shot up to the place of low pressure changes to gas taking in heat.
液体状態の冷媒は、圧力が低いところに噴出し、熱を奪いながら気体になる。
液體狀態的冷媒在壓力低的時候噴出，一面被帶走熱能、一面轉變為氣體。

Q68 How can you see yourself in a mirror?

①There are some things other than a mirror which can
reflect something. ② Some examples of these are windows
glass, still water, and so on.

③Things which reflect something have a characteristic in common:
their surfaces are even and smooth. ④Things which have this
characteristic show specular reflection of light. ⑤ Specular
reflection is a reflection in which light from a certain direction is
reflected at the equal angles.

⑥ A well-polished brass plate also shows specular reflection of light
but it does not reflect images as clearly as a mirror. ⑦ A brass
plate has a yellowish color because it takes in a bluish color, which
is complementary yellow.

⑧You may say that polished silver and aluminum have a silver
color, but in fact they are colorless. ⑨ Colorless means that they
reflect all colors in visible light in the same way. ⑩ A mirror, which
is made by putting a film of silver or aluminum on the back of
a pane of glass, is colorless.

⑪ It has the characteristic of showing specular reflection of light
without losing any light and you can see yourself in the mirror.

なぜ、鏡に姿が映るの？

答 ①鏡のほかにも、物体の像が映るものがあります。②たとえば、**窓ガラス、静か**
物體　　　　　　映る　　反射
な水面などです。
水面

③像が映りやすいものに**共通**しているのは、**表面が平ら**で、**ツルツルしている**ことで
共通　　　　　　　　表面
す。④こうした性質のものは光を「**正反射**」します。⑤正反射とは、**入射角と反射角**が
特徴　　　　　　　　鏡面反射　　　　　　　　　　　入射角　　反射角
等しくなる反射のことを指します。

⑥**ピカピカに磨き上げた真鍮板**も光を正反射しますが、**鏡ほどきれいに像を映しませ**
磨　　　　黄銅鏡
ん。⑦真鍮は、**黄っぽい色**をしていますが、これは**黄色の補色**である青っぽい色を吸
黄銅　　　　　　　　　　　　　　　　　　　黄色　補色　　　　　　　　　　　　吸収
収しているからです。

⑧磨いた**銀やアルミ**は「**銀色**」といわれていますが、実はどちらも**無色**です。⑨無色と
銀色　　　　　　　　　　　　　　　　無色
いうのは、**可視光線**に含まれているすべての色の光を、**同じように反射する**という意
可視光線　　　　　　　　　　　　　　　　　　　　反射　　　　　　　意指
味です。⑩**ガラスの裏側に銀やアルミの膜を貼った鏡は無色**なのです。⑪それは光を効
内側　　　　　　　　　貼　　　　　　　　　　　　　　　効率
率よく正反射する性質があるため、きれいに姿を映すのです。
鏡面反射　特性

為什麼鏡子會反射出影像？

答 ①除了鏡子之外，**還有其他的東西可以反射**物體的影像。②**舉例來說：窗戶的玻璃、靜止**
的水面…等等。
③容易反射影像的東西有一個共通的地方，就是**表面平坦、呈現光滑**的物品。④擁有這項特徵的
東西，可以進行光線的「**鏡面反射**」。⑤所謂的鏡面反射指的是光線在反射的時候，**入射角等於**
反射角。⑥**磨到閃閃發亮的黃銅鏡雖然**也可以進行光線的鏡面反射，但是卻**無法映照出像鏡子一**
般漂亮的影像。⑦黃銅呈現泛黃的顏色，是因為它吸收了黃色的補色—泛藍色所致。
⑧你會說拋光的銀或是鋁呈現「**銀色**」，事實上兩者**並無色彩**。⑨所謂的無色，指的是以**相同的**
方式，將所有的色彩反射成**可視光線**。⑩在**玻璃內側貼上**銀或鋁的薄膜的鏡子，是沒有顏色的。
⑪因為它有著能將光線作有效的鏡面反射的特性，所以能映照出漂亮的影像。

Q69　How does detergent remove dirt?

①It is difficult to remove greasy dirt on plates with only
　　　　　　　　　　去除　　　油汙　　　　　　碗盤
water. ②Oil and water naturally repel each other. ③
　　　　　　　　　　　　　　　　排斥
However, when you use detergent, you can wash away greasy dirt
　　　　　　　　　洗潔劑　　　　　　　洗淨
easily. ④This is because the major element of detergent, surfactant,
　　　　　　　　　　　主要成分　　　　　　　　　界面活性劑
is working. ⑤The molecules of the surfactant are like very small
　　　　　　　　分子
matchsticks. ⑥The heads of these "small matchsticks" are likely to
火柴棒　　　　　　　　　　　　　　　　　　　　　　　容易…
bind with water and the wood is likely to bind with oil. ⑦Thses
與…結合　　　　　　　　木頭（軸）
"small matchsticks" come between oil and water so that they can
　　　　　　　　　進入…當中　　　　　　　　　所以…
bind with each other easily.

⑧When you use detergent, the part of the surfactant which can
bind with oil first wraps greasy dirt in it. ⑨Many surfactants
　　　　　　　包、裹
come close to greasy dirt and they break it down into small pieces.
靠近…　　　　　　　　　　　　將…分解為…
⑩Next, water can come between the greasy dirt and the plate to
接下來
remove the greasy dirt from the plate because the outside of the
　　　　　　　　　　　　　　　　　　　　　　外側
surfactant is likely to bind with water. ⑪The function whereby the
　　　　　　　　　　　　　　　　　　機能
surfactant mixes oil and water uniformly in this way is called
　　　　　　　　　　　　　　平均地　　以這種方式
emulsification.
乳化（作用）

問69　洗剤はどうやって、汚れを落とすの？

答　①**お皿**についた**油汚れ**は、水洗いだけではなかなか落ちません。②油と水はなじ
　　　　油汚　　　　　水洗

みにくい性質があるためです。③しかし、**洗剤**を使うと、油汚れが水できれいに落ち
　　　　　性質　　　　　　　　　　　　　洗潔劑

ます。④これは、洗剤の**主成分**である**界面活性剤**という物質の働きによるためです。⑤
　　　　　　　　　　　　　　　　かいめん
　　　　　　　　　　　　　　　界面活性劑　　　　物質

界面活性剤の**分子**はとても小さい**マッチ棒**のような形をしています。⑥頭の部分は水
　　　　　　分子　　　　　　　　　　　　　　　　　　　　　　　　　部位

になじみやすく、棒の部分は油になじみやすい性質があります。⑦これが水と油の境
　　　　　　　　　　　　　　　　　　　　　　性質　　　　　　　　　　　　橋梁

目に入って両方をなじませます。
　　　　　雙方

⑧洗剤を使うと、まず、界面活性剤の油になじむ部分が油汚れ**を包み込みます**。⑨たく

さんの界面活性剤がくっついて油汚れを小さく**分解します**。⑩界面活性剤の**外側**は水
　　　　　　　　　　　　　　　　　　　　　分解　　　　　　　　　　　外側

になじみやすい部分なので、汚れとお皿の間に水が入り込んで汚れがお皿から離れて

いきます。⑪**このように**界面活性剤が、油と水が**均一**に混ざった状態にすることを**乳**
　　　　　　　　　　　　　　　　　　　　　　平均　　　　　　　　　　　　　　乳化

化〔**作用**〕といいます。
　　　作用

問69　洗潔劑是如何清潔髒汙的？

答　①沾在**碗盤**上的**油汙**，只用水洗是無法完全清洗乾淨的。②這是因為油水自然排斥的性質
　　　　所致。③但是，一使用**洗潔劑**，油汙就可以完全地洗淨。④這是因為洗潔劑中的**主要成**
分，稱之為**界面活性劑**的物質作用的緣故。⑤界面活性劑的**分子**，形狀有如相當細小的**火柴棒**。
⑥它有著頭部容易與水結合、棒狀部位容易與油結合的性質。⑦用這個做為油水的橋梁，讓雙方
結合在一起。⑧使用洗潔劑的時候，界面活性劑的親油部位會將油汙**包覆**。⑨大量的界面活性劑
會連在一起，將油汙**分解**成小塊。⑩而因為界面活性劑的**外側**是親水的部位，用水沖入油汙與碗
盤之間，油汙就會從碗盤上脫落。⑪**如此的**界面活性劑將油水**平均**混合的狀態，稱之為「**乳化作
用**」。

Why can erasers erase letters?

 ① Pencil lead is made by mixing black powder, black lead,
鉛筆筆芯　　　　　　將…混合　　　　黑鉛

with clay and other things and by firing it at a high
黏土　　　　　　　　　　　燃燒　　　用高溫

temperature. ② By rubbing it on paper, the powder of black lead
摩擦

sticks to the surface of the paper because of friction.
黏在…　　表面　　　　　　藉由摩擦

③ Some erasers for pencils are made of rubber and others are made
橡皮擦　　　　　　由…作成　橡皮

of plastic. ④ When you rub this powder of black lead with a rubber
橡膠　　　　　　摩擦…

eraser, the particles of the black lead on the paper stick to the
粒子

surface of the eraser. ⑤ When you keep on rubbing, these particles
持續地摩擦

and the surface of the eraser become residues. ⑥ Plastic erasers,
橡皮擦屑

which are softer than rubber ones, erase the powder of the black
去除…

lead on the paper by encasing it.
包覆

⑦ It is dificult to erase letters written with ballpoint pens with an
原子筆

eraser for pencils because the ink sinks into the paper. ⑧ In this
墨水　滲入…

case, we use sand erasers, which are made of rubber containing
磨砂橡皮擦　　　　　　　　　　　　　　包含

fine sands. ⑨ They can erase the letters written with ballpoint
細小的

pens by chipping off the surface of the paper which the ink sinks
削除…

into.

問70　どうして、消しゴムで字が消せるの？

答 ①鉛筆の芯は**黒鉛**という黒い粉を**粘土**などと混ぜ、**高温で焼いた**ものです。②紙にこすることにより、**摩擦**で黒鉛の粉が紙の**表面**に付着します。

③鉛筆用の**消しゴム**には、**ゴム製**のものと**プラスチック製**のものがあります。④ゴム製の消しゴムでこの黒鉛の粉**をこする**と、紙の表面の黒鉛の**粒子**がゴムの表面に付着します。⑤さらにこすると、これがゴムの表面とともに消し**カス**となります。⑥ゴム製のものよりも柔らかいプラスチック製の消しゴムは、紙の表面の黒鉛の粉を**包み込む**ようにして取り除きます。

⑦ボールペンで書いた**文字**は、**インク**が紙に**しみ込んで**しまっているので、鉛筆用の消しゴムではなかなか消えません。⑧そこで、ゴムに**細かい砂**を混ぜてつくった「**砂消し**」を使います。⑨インクがしみ込んでいる紙の表面を砂で**削り取る**ことで、ボールペンの字を消すことができるのです。

問70　為什麼橡皮擦可以擦掉字跡？

答 ①**鉛筆的筆芯**是由稱為**黑鉛**的黑色粉末**混合黏土**，經由**高溫燒製**而成。②藉由在紙上**摩擦**，將黑鉛的粉末附著於紙的**表面**。③鉛筆用的**橡皮擦**，有分為**橡皮製**與**橡膠製**。④用橡皮製的橡皮擦**摩擦**這個黑鉛的粉末時，紙的表面的黑鉛粒子會附著於橡皮擦的表面。⑤再接著摩擦時，這個粉末會跟橡皮的表面一起形成橡皮擦屑。⑥比起橡皮製還要柔軟的橡膠製橡皮擦，將紙的表面黑鉛以**包覆**的方式去除。

⑦用**原子筆**書寫的文字，由於墨水滲入紙張內，用鉛筆用的橡皮擦是無法擦拭乾淨的。⑧這個時候，就要使用在橡皮裡摻入**細砂**混合的「**磨砂橡皮擦**」。⑨將墨水滲入表面的紙張**用磨砂橡皮擦**削除表面，就能夠去除原子筆的筆跡。

Q71　Why can an iron smooth wrinkles?

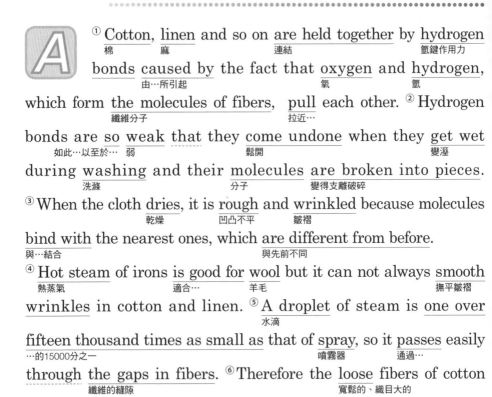

①Cotton, linen and so on are held together by hydrogen
（棉）（麻）（連結）（氫鍵作用力）
bonds caused by the fact that oxygen and hydrogen,
（由…所引起）（氧）（氫）
which form the molecules of fibers, pull each other. ②Hydrogen
（纖維分子）（拉近…）
bonds are so weak that they come undone when they get wet
（如此…以至於…）（弱）（鬆開）（變溼）
during washing and their molecules are broken into pieces.
（洗滌）（分子）（變得支離破碎）
③When the cloth dries, it is rough and wrinkled because molecules
（乾燥）（凹凸不平）（皺褶）
bind with the nearest ones, which are different from before.
（與…結合）（與先前不同）
④Hot steam of irons is good for wool but it can not always smooth
（熱蒸氣）（適合…）（羊毛）（撫平皺褶）
wrinkles in cotton and linen. ⑤A droplet of steam is one over
（水滴）
fifteen thousand times as small as that of spray, so it passes easily
（…的15000分之一）（噴霧器）（通過…）
through the gaps in fibers. ⑥Therefore the loose fibers of cotton
（纖維的縫隙）（寬鬆的、織目大的）
and linen do not hardly get wet.
（幾乎不會）
⑦By moistening these fabrics not with the steam of irons but with
（弄溼…）（不使用…）
spray, you can easily smooth wrinkles in cotton and linen .

⑧Bound molecules come undone when you moisten the cloth.
（結合的）
⑨Ironing at that time makes water evaporate and the molecules
（熨燙時）（蒸發）
bind with one another again without wrinkles because of the
（交互地）（壓力）
pressure.

問 71　アイロンをかけると、なぜシワが伸びるの？

答 ①綿や麻などは、**繊維の分子を構成する**酸素と水素が引き合うことから生まれる**水素結合で結びついています。**②水素結合は弱いので、**洗濯して水分を含むことにより解け、分子がばらばらになります。**③水分が乾燥するときに、元の結合とは関係なく近くにある分子同士が結合するので、**凸凹になり、シワができます。**

④**ウール**には効果的なアイロンの**スチーム**機能ですが、綿や麻だと**シワがよく伸びない**ことがあります。⑤スチームの**水滴**は、**霧吹きの水滴の15000分の1**というごく小さいもので、**繊維の隙間をたやすく通過してしまいます。**⑥綿や麻など**目の粗い繊維**だとほとんど水分が残らない**のです。

⑦アイロンのスチーム機能を使わず、霧吹きで水分を与えてやれば、綿や麻でもシワをうまく伸ばすことができます。⑧水分を与えることにより、**結合した分子がゆるみます。**⑨そこに**アイロンをかけることにより**水分が気化し、**圧力**によりシワのない状態で分子が**再結合す**るのです。

問 71　為什麼用熨斗可以燙平皺褶？

答 ①棉和麻等其他布料，是經由構成纖維分子的氧和氫結合過程中產生的**氫鍵作用力**連結在一起。②由於氫鍵作用力相當**脆弱**，會經由**洗滌溶解在水中，分子變得支離破碎。**③當水分乾燥時，便就近與原本不一樣的相同分子結合，因此會形成**凹凸**或皺褶。
④熨斗的蒸氣功能對羊毛製品有效，但是卻不能**撫平棉或麻製品的**皺褶。⑤由於蒸氣**水滴**是噴霧器水滴的**15000分之一**，相當的微小，因此能**輕易地通過纖維的縫隙**。⑥像棉或麻這類織目大的纖維，**幾乎不會**殘留水分。
⑦若不使用熨斗的蒸氣功能，用噴霧器給與水分的話，即使棉或麻的皺褶也能輕易地撫平。⑧因為給予水分的時候，**結合的分子會鬆弛。**⑨這時候用熨斗將水分蒸發，就能用**壓力**將分子結合成平坦的狀態。

Detergent removes dirt because it can bind with both water and oil.

洗剤が汚れを落とすのは、水にも油にもなじむから
洗潔劑能夠除去髒污，是因為它能夠讓油水結合

The molecules of the surfactant agent in detergent

洗剤に含まれる界面活性剤の分子
洗潔劑中含有介面活性劑的分子

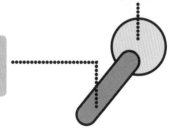

The part which is likely to bind with water
水になじみやすい部分
容易與水結合的部分

The part which is likely to bind with oil
油になじみやすい部分
容易與油結合的部分

This is how dirt is removed.

汚れが落ちるしくみ
如何洗淨髒污

1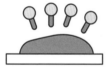

The part of the surface surfactant which is likely to bind with oil faces to the surface of dirt.

界面活性剤が、油になじみやすい部分を汚れの表面にむけて集まる
界面活性劑容易與油結合的部分，朝向髒污的表面集結

2

The surfactant acting agent wraps greasy dirt in it.

界面活性剤が汚れを包み込む
界面活性劑將髒污包覆

3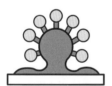

Taking dirt away into water little by little.

少しずつ汚れを水中に取り出す
一點一點地將髒污融入水中

4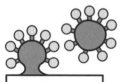

Dirt is removed by flushing with water.

水ですすぐと、汚れは洗い流される
用水沖洗後，髒污就被清洗乾淨

The molecules of fibers line up by ironing after moistening with spray.

霧を吹いた後、アイロンをかけると、繊維の分子がきれいに並ぶ

噴霧器噴過後，用熨斗燙過，纖維的分子就能整齊地撫平

The fibers which are rough and wrinkled

凸凹になり、シワが寄った繊維

凹凸不平、有皺褶的纖維

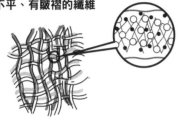

Hydrogen and oxygen bind with each other disorderly.

水素と酸素が不規則に結合している

氫分子和氧分子不規則地結合在一起

This is how wrinkles are smoothed.

シワが伸びるしくみ

如何撫平皺褶

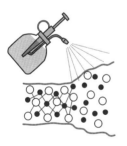

When you moistening by spray, disorderly bound molecules come undone.

霧吹きで水分を与えると、不規則な分子の結合が解ける

一用噴霧器灑上水分，就能讓不規則的分子結合散開。

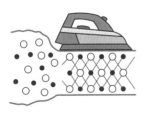

Smooth fibers which are not rough or wrinkled any more

凸凹がなくなり、シワが伸びた繊維

去除凹凸，撫平皺褶的纖維

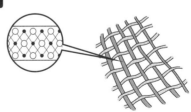

The molecules line up.

分子がきれいに並んでいる

分子整齊地排列在一起

Q72　Why does a magnet attract iron?

① A magnet attracts metals such as iron, nickel, cobalt,
磁鐵　　　吸引　　　金屬　　　　　　鐵　　 鎳　　　鈷
and so on. ② A magnet attracts them because these

metals themselves became a magnet for a time by placing a
　　　他們自己　　　　　　　　　　　　暫時地　　　　放置
magnet close to them.

③ As you know from the fact that iron becomes an electromagnet
當…　　　　　　　事實是…　　　　　　　　　電磁鐵
when you wind nichrome wire around the iron, a magnet has
　　　　捲　　 鎳鉻線圈
something to do with rotation. ④ In the molecules of metal,
與…有關　　　　　　旋轉　　　　　分子
electrons rotate around an electron-shell at the center. ⑤ A
電子　　繞著…旋轉　　　電子核
magnetic force is produced by these rotations of the electrons.
磁力
⑥ However, in the case of most of the metals, electrons have their

pair rotating in the opposite direction and the pairs cancel out the
一對　　　　　反方向的　　　　　　　　　　　取消…
magnetic force of each other. ⑦ In the case of iron and other metals,

there are electrons which have no pairs, so a magnetic force is

produced.

⑧ However, iron is not usually a magnet. ⑨ Imagine that there are
　　　　　　　　普通　　　　　　　　　假設…
countless tiny magnets in iron. ⑩ The countless magnets are facing
無數個　 小　　　　　　　　　　　　　　　　　　　　朝向
in different directions and cancel out the magnetic force of one
各個方向
another. ⑪ By placing a magnet close to them, they face in the
　　　　　　　　　　　　　　　　　　　　　　 朝著同一個方向
same direction and the iron becomes a magnet temporarily.
　　　　　　　　　　　　　　　　　　　　 暫時地

問72　なぜ、鉄と磁石はくっつくの？

答 ①磁石にくっつく金属は、鉄、ニッケル、コバルトなどです。②磁石を近づける
　　　　磁鐵　　　　　　　金属

ことにより、これらの金属自体が一時的に磁石になるため、磁石にくっつくのです。
　　　　　　　　金屬本身　暫時　磁鐵

③鉄にニクロム線を巻き、電流を流すと電磁石になることからもわかるように、磁石
　　　　　　　　　　　電流

は回転と関係があります。④金属の分子の中では、電子核を中心に電子が回っていま
　旋轉　關係　　　　　　　金屬　　　　　　電子核　中心

す。⑤この電子の回転によって、磁力が生まれます。⑥ただし、ほとんどの金属では、
　　　　　　旋轉　　　　　磁力

逆回りの電子が対になっていて、磁力を打ち消し合います。⑦鉄などでは、対になっ
相反旋轉

ていない電子があるため、磁力が生まれるのです。

⑧しかし、鉄は普通の状態では磁石ではありません。⑨鉄の中に小さな磁石が無数にあ
　　　　　　普通　　　　　磁鐵　　　　　　　　　　　　　　　　　　　　無數

ると考えてみましょう。⑩無数の磁石がばらばらの方向をむき、磁力を打ち消し合っ

ています。⑪磁石を近づけることにより、その向きがそろって一時的に磁石になるの
　　　　　　　　　　　　　　　　　　　　　　　　　暫時地　磁鐵

です。

問72　為什麼磁鐵會吸附在鐵上？

答 ①磁鐵會吸附的金屬有鐵、鎳、鈷等等。②將磁鐵靠近時，由於這些金屬本身會暫時變成磁鐵，因此會吸附磁鐵。③如同我們已知將鐵塊捲上鎳鉻線圈，通電之後會成為電磁鐵一般，磁鐵與旋轉有關。④金屬的分子當中，電子以電子核為中心圍繞著旋轉。⑤因為這個電子的旋轉，而產生了磁力。⑥但是，幾乎所有的金屬有著相反的電子在旋轉相抗，抵消了磁力。⑦而鐵等等金屬，因為沒有相反的電子轉相抗，所以產生磁力。

⑧但是，鐵在普通的狀態下並不是磁鐵。⑨我們假設在鐵之中有無數個小磁鐵存在其中。⑩無數的磁鐵散落在各個方向，因而抵消了磁力。⑪當磁鐵靠近的時候，會朝著同一個方向聚集，所以暫時地變成了磁鐵。

Q73 Why does it become white when light's three primary colors are mixed?

①Visible light, which human beings are able to see, has
可視光線　　　　　　　　　　　　　　　　可以

many colors in it. ②They are the same colors as the
　　　　　　　　　　　　　　　　　　　與…相同

many colors from red to purple in the rainbow, which is called the
　　　　　　　　　　紫　　　　彩虹

prism in nature.
三菱鏡

③It is human beings that feel colors. ④The retina of the eye has
據說…　　　　　　　　　　　　　　　　　視網膜

three kinds of cone cells: L, M, and S. ⑤Red light excites L cones,
　　　　錐狀細胞　　　　　　　　　　　　　　讓…興奮　L錐狀細胞

green light excites M cones, and blue light excites S cones.

⑥Human beings can tell colors apart by these three kinds of cone
　　　　　　　　辨別

cells. ⑦When these cone cells get a well-balanced stimulus, human
　　　　　　　　　　　　　　　　平均的刺激

beings feel it is white.

⑧In other words, the fact is not that it becomes white when light's
換言之…　　　　事實並分如此　　　　　　　　　　　　　光的三原色

three primary colors are mixed but that the color which is made
　　　　　　　　　　混合　　　　　　　　　　　　　　由…所組成

from light's three primary colors is called white. ⑨However, the

color white is neutral and is special for human beings. ⑩This is
　　　　　　　中性

because white is the most commonly seen color and it is the color
　　　　　　　　　最普遍的

similar to direct sunlight.
　　　　　直射陽光

問73 光の三原色を合わせると、なぜ白くなるの？

答 ①人間の眼で見ることの**できる可視光線**には、いろいろな色の光が混ざっています。
人類　　　　　　　　　　　　　　　　可視光線
②自然の**プリズム**といわれる虹の中に見える赤から**紫**までの色がこれにあたります。
自然界

③色を感じとるのは、人間です。④眼の**網膜**にはL、M、Sの３種類の**錐体細胞**があり
人類　　　　　　視網膜　　　　　　　　　　　　　　　　錐狀細胞
ます。⑤赤い光は**L錐体**を、緑の光はM錐体を、青い光はS錐体を刺激します。⑥これ
刺激
ら３種類の錐体細胞によって、人間は色**を見分ける**ことができるのです。⑦そして、
種類　　　　　　　　　　　人類
これらが**バランスよく刺激される**と、人間は白であると感じます。
刺激
⑧つまり、「光の三原色を合わせると白くなる」の**ではなく**、「**光の三原色を合わせ**
三原色
た色を白としている」の**です**。⑨にもかかわらず、私たちにとって白は**ニュートラル**
で特別な色です。⑩なぜなら、白は私たちが一番見慣れた、**直射日光**に近い色だから
特別　　　　　　　　　　　　　最常見　　　　直射陽光
です。

問73 為什麼將光的三原色混合之後，會形成白色？

答 ①人類的眼睛可以見到的**可視光線**之中，混雜著各式各樣的顏色。②這些顏色與被稱為自然界的三菱鏡—彩虹中所見的紅、橙、黃、綠、藍、靛、**紫**等顏色相同。③能夠察覺得到顏色的，只有人類。④人眼中的**視網膜**裡，含有L、M、S三種類的**錐狀細胞**。⑤紅色光刺激L錐狀細胞、綠色光刺激M錐狀細胞、藍色光刺激S錐狀細胞。⑥因為這三種錐狀細胞，人類可以**區分**出顏色。⑦而且，當這些錐狀細胞受到**平均的刺激**時，人類會感覺到白色。
⑧也就是說，不是「混合光的三原色就會變成白色」，**而是**「**由光的三原色所組成的顏色呈現白色**」。⑨可是對我們來說，白色是**中性的**特別色。⑩這是因為白色是我們最常見、最接近**直射陽光**的顏色。

Q74　Why don't you get burned in a sauna bath?

① The temperature in a dry sauna bath is very high,
温度　　　　　　　　　乾式三溫暖
around 80 to 100 degrees C. ② If you put your finger in
80～100度以上　　　　　　　　　　放入…
boiling water at a temperature higher than 90 degrees C, you will
熱水　　　　　　　　　　90度以上
get badly burned in no time.
被…（燙傷）　　　立刻

③ However, why don't you get burned in a hot sauna bath?

④ The first reason is that it is dry in a sauna bath. ⑤ When your
乾燥
body gets hot, you sweat. ⑥ In a sauna bath, sweat exudes from
變熱　　　　　　　　流汗　　　　　　　　　　　散發
your body because it is dry. ⑦ When this happens, vaporization
這個時候　　　　汽化熱
heat is needed and the heat of your body is used for it and your
body will be covered with cool a layer of air.
被…覆蓋　　　　　層
⑧ The second reason is that air doesn't transmit heat as well as
傳導　　　　　　和…一樣
water. ⑨ When you put your finger in boiling water, its heat is
transmitted to your finger directly and you will get burned.
直接
⑩ However, heat is not transmitted to your body directly when you
take a hot sauna bath naked. ⑪ The cool layer of air covering your
洗熱的三溫暖　　　　　裸體
body cuts off heat to some extent. ⑫ You feel pricking heat when
阻絕　　　　某種程度上　　　　　　　刺痛
you walk around in a sauna bath because the layer of air partly
部分被破壞
breaks down. ⑬ For these two reasons, you can take a sauna bath
at higher than 90 degrees C for a long time without getting
burned.

問 74　なぜ、サウナでヤケドしないの？

答 ① 乾式のサウナの中の**気温**は**80～100度以上**の高温です。② もしも**90度以上の熱湯**に
　　乾式　　　　　　　　氣溫　　　　　　　　　　　　高溫　　　　　　　　　　熱水

指を入れたら、たちまち大やけどをしてしまいます。③ それなのに、熱いサウナに入っても**ヤケド**

しないのは、なぜでしょうか？

④ 第1の理由は、サウナ内部の湿度が低いことです。⑤ 体が**熱くなる**と、人間は**汗をかきます**。
　　　　理由　　　　　　内部　　　　　　　　　　　　　　　　　　　　　人類

⑥ 湿度が低いため、汗は体の表面で**蒸発します**。⑦ このとき、**気化熱**として体の表面の熱を奪
　　　　　　　　　　　　　　　　蒸發

い、同時に体の表面に温度が低い空気の**層**ができます。
　　同時　　　　　　　　　　　　空氣

⑧ 第2の理由は、空気は水に比べて熱**を伝え**にくいことです。⑨ お湯に指を入れると、お湯の熱

が**直接**指に伝わり、ヤケドします。⑩ ところが、高温の**サウナに裸で入っても**、熱は体に直接伝
　　直接　　　　　　　　　　　　　　　　　　高溫

わりません。⑪ 体の周りにできた、温度が低い空気の層も、熱を**ある程度遮断して**くれます。⑫
　　　　　　　　　　　　　　　　　　　　　　　　　　　　　　　程度 阻絕

サウナの中で動くと、**チクチクする**ような熱さを感じるのは、空気の層が乱れるからです。

⑬ これら二つの理由により、人間は90度以上もあるサウナに長時間入っていても、ヤケドする

ことがないのです。

問 74　為什麼在三溫暖裡不會燙傷？

答 ① 乾式三溫暖當中的氣溫為80～100度以上的高溫。② 假如將手指伸入90度以上的熱水中，

　　立刻就會嚴重燙傷。③ 如果是這樣，那麼進入熱的三溫暖卻**不會燙傷**，這是為什麼呢？
④ 第1個理由是，三溫暖內部的溼度很低。⑤ 身體一**變熱**，人類就**開始流汗**。⑥ 由於濕度低的緣
故，汗在身體的表面蒸發。⑦ 這個時候由於**汽化熱**的作用，奪走了身體表面的熱度，同時在身體
表面形成低溫的空氣層。
⑧ 第2個理由是，空氣與水相較之下並不容易**傳導**熱度。⑨ 將手指一伸入熱水中，熱水的熱度會**直
接**傳到手指，造成燙傷。⑩ 可是，若是**裸體進入高溫**的三**溫暖**，熱度不會直接傳到身體上。⑪ 在
身體周圍形成的低溫空氣層，也能在**某種程度上阻絕**熱度。⑫ 在三溫暖中移動時，會感覺到猶如
陣陣刺痛的熱度感覺，那是因為部分空氣層被破壞的緣故。⑬ 因為這兩個理由，即使人類長時間
進入90度以上的高溫三溫暖，也不會燙傷。

Q75 Why do clock hands move to the right?

① Clock hands move to the right because clocks were
時鐘的指針　　　向右邊旋轉　　　　　　　　被發明
invented in the Northern hemisphere.
北半球

② There used to be various clocks such as water clocks, sand
以前曾經有…　　　　　　　　　　　　水時鐘　　　砂時鐘
clocks, sun clocks, and so on before mechanical clocks, which have
　　　日晷　　　　　　　　　　機械時鐘
moving hands, began to be made.

③ Among them, sun clocks were very popular. ④ Sun clocks tell the
　　　　　　　　　　　　　　　　　　　　　　　　　報時
hour by the shadow the sun makes. ⑤ It is thought that the shadow
　　　　　　　陽光的影子
moves to the right in the Northern hemisphere, so mechanical
clocks were first made to move to the right.

⑥ Not only mechanical clocks but also almost all meter hands
不只…　　　　　　　　　　　而且…　　　　　　計量器的指針
move to the right. ⑦ It is true that they were made to move to the
　　　　　　　　　事實上…
right like a clock but moving to the right like the movement of the
sun can be natural for the people living in the Northern hemisphere.
有可能…
⑧ There are quite a few left-handed clocks in the world such as the
　　　少數的　　向左旋轉的時鐘
reverse clock in a barber shop, in which clients see the clock in a
反向時鐘　　　理髮廳　　　　　　　顧客
mirror, and watches for special use.
映照在鏡子裡　　　特別的用途
⑨ However, people have difficulty accepting them partly because
　　　　　　　　　　　　　　　　　　　　　　不完全地
they are used to clocks moving to the right in daily life and think
習慣…　　　　　　　　　　日常生活中
it is strange.

問75　時計の針はどうして、右回りなの？

答 ①時計が**右回り**なのは、時計は**北半球**で**発明された**からです。②現在のように針
時鐘　　　　　　　　　　　　　　北半球　發明　　　　　　　　　現代

が回転する**機械時計**がつくられるようになる前には、**水時計**、**砂時計**、**日時計**といっ
機械時鐘　　　　　　　　　　　　　　　　　　水時鐘　砂時鐘　日晷

た、さまざまな時計がありました。③なかでも広く使われていた日時計です。④日時計

は**太陽の影**が時間を示します。⑤北半球では影は右回りに動くので、それに合わせて

機械時計も右回りにしたのでしょう。
機械時鐘

⑥時計に限らず、ほとんどすべての**計器の針**は右回りです。⑦時計に合わせて右回りに
計器

したというだけでなく、北半球に暮らす私たちにとって、太陽の移動と同じ右回りの
移動

ほうが、左回りよりも自然なのかもしれません。

⑧**理髪店**が**鏡に映してみる**ためにつくられた**逆回転時計**や特殊な**腕時計**など、左回り
理髮廳　　　　　　　　　　　　　　　　　反向時鐘　　特殊　手錶

の時計も存在しないことはありません。⑨しかし、常日頃から**右回りに慣れており**、
日常　　　　　　右邊開始旋轉

違和感があるためか、ほとんど受け入れられないようです。
不協調

問75　時鐘的指針為什麼是往右旋轉呢？

答 ①時鐘向**右邊旋轉**是因為，時鐘是在**北半球發明**出來的。②在製作出現代這種指針旋轉的**機械時鐘**以前，還有**水時鐘**、砂時鐘、日晷等各式各樣的時鐘。③當中被廣泛使用的是日晷。④日晷是由**陽光的影子**來表示時間。⑤在北半球影子是往右邊旋轉移動，因為配合它所以機械時鐘也設計成往右邊旋轉。⑥不只是時鐘，幾乎所有的**計量器的指針**都是往右邊旋轉。⑦不只配合時鐘往右旋轉，對於生活在北半球的我們來說，跟著太陽移動向右邊旋轉，比起向左邊旋轉說不定還比較自然。⑧世界上也存在著類似為了能**映照在理髮廳的鏡子**，而製作的**反向時鐘**或特殊的**手錶**等等向左旋轉的時鐘。⑨但是，日常已經**習慣**了由右邊開始旋轉，因為會產生不協調的緣故，幾乎不容易被接受。

When tiny magnets in iron face in the same direction, the iron becomes a magnet.

鉄の中の小さな磁石の向きがそろうと、鉄は磁石になる
當磁鐵中的小磁石朝向同一個方向，鐵塊就變成磁鐵

Iron
鉄
鐵

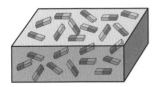

There are tiny magnets in iron and they are usually facing in different directions.

鉄の中には小さな磁石があり、通常はばらばらの方向をむいている。

鐵塊之中含有小塊的磁石，通常散亂地朝著不同的方向。

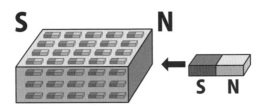

When you move a magnet close to them, they face in the same direction and the iron becomes a magnet.

磁石を近づけると小さな磁石の向きがそろい、鉄が磁石になる。

當磁鐵靠近的時候，小塊的磁石會朝向相同的方向，鐵塊就變成了磁鐵。

Electromagnet
電磁石
電磁鐵

Iron bar
鉄の棒
鐵棒

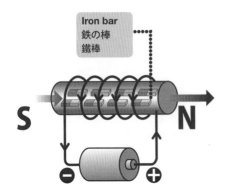

In an electromagnet, tiny magnets face in the same direction by a flow of electricity.

電磁石は、電気の流れにより、小さな磁石の向きがそろう。

電磁鐵是藉由電流的流動，讓小塊的磁石朝向同一個方向。

You do not get burned in a hot sauna bath because it is dry.

サウナでヤケドしないのは、湿度が低いから
在三溫暖中不會燙傷的原因，是因為濕度低的緣故

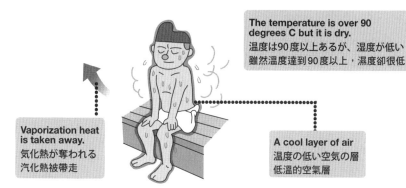

The temperature is over 90 degrees C but it is dry.
温度は90度以上あるが、湿度が低い
雖然溫度達到90度以上，濕度卻很低

Vaporization heat is taken away.
気化熱が奪われる
汽化熱被帶走

A cool layer of air
温度の低い空気の層
低溫的空氣層

Sweat exudes much from your body because it is dry in a sauna bath and the heat on the surface of your body is taken away. At the same time, a cool layer of air covering your body cuts off heat to some extent.

サウナの中は湿度が低いため、汗が盛んに蒸発し、それが体の表面の熱を奪う。同時に冷たい空気が体表に層をつくり、その層が熱をある程度遮断してくれる。

由於三溫暖中的濕度低，大量蒸發的汗帶走身體表面的熱度。同時冷空氣在體表形成空氣層，這層空氣層能夠在某種程度上隔絕熱度。

Clock was first made to move to the right because the shadow of the sun clock moves to the right.

時計の針が右回りなのは、日時計が右回りだから
時鐘的指針朝向右邊轉動，是因為日晷向右邊旋轉的緣故

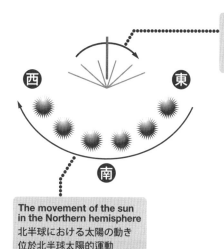

The movement of the shadow of the sun clock in the Northern hemisphere
北半球における日時計の影の動き
位於北半球的日晷的影子的移動

西　東

南

The movement of the sun in the Northern hemisphere
北半球における太陽の動き
位於北半球太陽的運動

The sun moves to the right in the Northern hemisphere, so clock was first made to move to the right.
北半球では、太陽の移動が右回りに動くので、時計の針も右回りになった
因為在北半球太陽的移動是向右旋轉，因此時鐘的指針也向右旋轉

Q76 Why does a playback of your recorded voice sound strange?

①When someone else is speaking, his or her real voice 自然的嗓音 and a playback of the recorded voice sound similar. ② 播放　　　　　錄音的聲音　　　聽起來　相似的 However, when you record your voice and play it back, the 播放 recorded voice sounds different. ③Why is this?

④A playback of the recorded voice travels through air to your ear, 透過空氣傳遞 and then it is passed through the outer ear, the middle ear, and 被傳遞到　　　　　　外耳　　　　中耳 the inner ear finally reaching the auditory nervous center in the 內耳　　　　　　　　　聽覺神經中樞　　　　　　　　大腦 brain . ⑤This is called air-transmitted sound. 氣導音 ⑥When you hear your own voice, bone-transmitted sound, which 骨導音 travels through the vocal cords and the skull bone to the auditory 聲帶　　　　　　　頭蓋骨 nervous center, and air-transmitted sounds are mixed. ⑦So your 混合的 real voice and a playback of your recorded voice sound different because of the difference in the way these sounds travel. 這些聲音的傳導方式 ⑧The sound from a recorder travels as a air-transmitted sound. 錄音機 ⑨Therefore a playback of your recorded voice sounds more like 像… your voice that others usually listen to. 平常他人聽到的 ⑩However, the recorded voice changes a little depending on the 根據… kind of microphone and player. ⑪It is not easy to listen to your …的種類　麥克風 own voice as others listen to it. 像…一樣

問76　どうして、録音した自分の声は変なの？

答　①他人がしゃべっている場合は、**肉声**と録音後に**再生**した声は同じように**聞こえ**
　　　　　時候　　自然の嗓音　　録音後　　再播放
ます。②自分の声を録音すると、生の声と違って聞こえます。③これは、なぜでしょう

か？

④再生された声は、**空気を伝わって**耳に入り、**外耳**から**中耳**、**内耳**に伝わり、**大脳**の
　再播放　　　　　　空氣　　　　　　外耳　　中耳　　内耳　　　　大脳
聴覚神経中枢へたどり着きます。⑤これは**気導音**と呼ばれています。
聴覚神経　　　　　　　　　　　　　　　氣導音
⑥自分の声は、**声帯**から**頭蓋骨**に伝わり聴覚神経中枢にたどり着く**骨伝導音**と、**気導**
　　　　　　　声帯　　頭蓋骨　　　　　聴覚神経　　　　　　　　骨導音　　氣導音
音がミックスされたものです。⑦**このような音の伝わり方**の違いから、自分の生の声

と再生された声は、違うように聞こえるのです。

⑧**レコーダー**の音は、気導音として伝わります。⑨したがって、レコーダーで再生され

たあなたの声のほうが、**いつも他人が聞いている**あなたの声に近いことになります。
　　　　　　　　　　　　　　他人
⑩しかし、**マイクの種類**や再生装置などによって、声が微妙に変わります。⑪他人が聞
　　　　　　録音装置　　　　　　　　　　　　　　微妙
いているのと同じように自分自身の声を聞くのは、簡単ではないのです。
　　　　　　　　　　　自己　　　　　　　　　　　　容易

問76　為什麼自己錄音過後的聲音會改變？

答　①別人在說話的時候，**自然的嗓音**跟錄音後**再播放**的聲音聽起來是一樣的。②錄自己的聲音時，聽起來就跟自然的嗓音不同。③這是為什麼呢？④播放的聲音透過空氣傳入耳朵，由**外耳**經由**中耳**傳遞到**內耳**，最終抵達大腦的**聽覺神經中樞**。⑤這就被稱之為氣導音。
⑥跟自己的聲音，由**聲帶**傳到**頭蓋骨**，最終抵達聽覺神經中樞的**骨導音**相比，氣導音屬於**混合音**。⑦因為**聲音傳導的方式**不同，自己的自然嗓音與錄音後播放的聲音，聽起來就會不同。
⑧**錄音機的聲音**也是以氣導音的方式傳遞。⑨因此，用錄音機播放你的聲音的時候，與**平常他人聽到**的你的聲音是相近的。⑩但是，根據**麥克風種類**與錄音裝置等的**不同**，聲音會有微妙的改變。⑪跟他人聽聲音的時候相同，要聽到自己的聲音也不是這麼容易。

Q77 How are cell-phones connected?

① The cell-phone network is made up of cell-phones you
行動電話網絡　　　　　　由…組成　　　　　　行動電話

have, base stations, the cell-phone switching station, and
基地台　　　　　　行動電話交換中心

the lines between them. ② The base stations are equipment such as
迴路　　　　　　　　　　　　　　　　　　　設備

antennas which connect to cell-phones by air and they are
天線　　　　連結　　　　　　　　　藉由無線

installed on the roofs of apartments or other buildings in your
被設置在…

town. ③ The base stations are connected to cell-phone switching

stations by optical cable and similar devices.
　　　　　光纖網路　　　同樣的設備

④ Imagine that you turn on your cell-phone. ⑤ It connects to a base
假設…　　　　　　打開

station and lets the cell-phone switching station know which base
讓…知道　　　　　　　　　　　　　哪一個

station's area it is in. ⑥ The cell-phone switching stations record
紀錄…

the position information of your cell-phone and which base station's
位置資訊

area your cell-phone is in. ⑦ When someone calls you, the cell-
打電話給

phone switching station finds the nearest base station using the
使用…

position information of your cell-phone and calls your cell-phone by
呼叫…

radio waves. ⑧ Even if you move to another place within the area
電波　　　　即使…　　　　　　　　　　　　在…的範圍內

of the cell-phone network, you can talk on the cell-phone because

your cell-phone connects to the base station by itself to update the
　　　　　　　　　　　　　　　　　　　　　　自動地　　更新

position information in the cell-phone switching station. (see p.204)

問77　携帯電話はどうして、つながるの？

答 ①携帯電話は私たちの手元にある携帯電話機のほか、**基地局**、**交換局**、それらを
　　　　行動電話　　　　　　　　　手邊　　　　　　　　　基地台　行動電話交換中心

結ぶ**回線**によって成り立っています。②基地局というのは電話機と**無線**でやりとりす
　　　回路　　　　　　　　　　　　　　　　　　　　　　　　電話機　無線

るための**アンテナ**を中心とする設備で、マンションやビルなどの屋上に**設置**されてい
　　　　　　　　　　　　　　　　　　　　　　　　　　　　　屋頂　　設置

ます。③それらは**光ケーブル**などにより、交換局と結ばれています。

④あなたが、携帯電話の**スイッチを入れた**としましょう。⑤電話機は基地局とやりとり

して、どの基地局にいるのかを交換局に伝えます。⑥交換局はあなたの携帯電話がど

の基地局のエリアにいるかを、**位置情報**として**記録**します。⑦あなたに**電話がかかっ**
　　　　　　　　　　　　　　　　資訊　　　　　紀錄

てきた場合、交換局は位置情報から最寄りの基地局を探し、**電波**であなたの**携帯電話**
　　　　　　　　　　　　　　　　最近　　　　　　　　　　電波　　　　　　　行動電話

を呼び出します。⑧別の場所に移動しても、あなたの携帯電話が**自動的に**基地局とや

りとりして交換局の位置情報を**更新**するので、携帯電話の**エリア内**であれば、どこに
　　　　　　　　　位置資訊　　更新

行っても通話することができるのです。

問77　行動電話是如何連結的？

答 ①行動電話除了我們手邊的行動電話機之外，還要連接**基地台**、**行動電話交換中心**，才能
組織成一個迴路。②所謂的基地台是為了電話機用**無線**的方式**交換訊號**，以天線為中心的
設備，**設置**在大樓或公寓等的屋頂。③這些設備經由**光纖網路**等，與行動電話交換中心連結。④
請你把行動電話的**電源開啟**吧！⑤行動電話機與基地台連結，並且將所在的基地台位置，傳送到
行動電話交換中心。⑥行動電話交換中心將你的行動電話所在的基地台位置，以**位置資訊**的方式
紀錄下來。⑦當你在**接電話**的時候，行動電話交換中心就由位置資訊當中搜尋最近的基地台，並
以**電波**的方式**呼叫**你的行動電話。⑧即使移動到別的場所，因為你的行動電話會**自動地**與基地台
連結，**更新**行動電話交換中心的位置資訊，只要是在行動電話的**收訊範圍**內，不論走到哪裡都可
以進行通話了。

Q78 How can airplanes fly?

①When you hold a plastic board with sloping upwards to
　　　　　拿著　塑膠板　　　　　　　　以前方朝上的方式傾斜

its front end and you run, you feel the power of the

plastic board to try to move upwards. ②This mechanism is similar
　　　　　　　　　　　　　　　　　　　　　　　　機制　　　　　　跟…相似

to that of flying airplanes.
　　　　　　飛機

③Whey an airplane moves forward, its wings catch the wind.
　　　　　　　　　　前進　　　　　　　　機翼　接受

④The wind which passed by the wings turns down a little. ⑤As a
　　　　　　　　通過　　　　　　　　　　向下改變　　　　　　　　結果

result, it puts upward force on the wings.
　　　　施加…　向上的力量

⑥This is caused as a reaction force of downward flow of air. ⑦It
　　　　由…引起　　反作用力　　　　　向下的空氣流動

becomes a lifting power to lift up the whole airplane.
　　　　　揚力　　　　　　　抬起…　機體

⑧The wings of the jet airliners today is thicker than before and is
　　　　　　　　噴射客機　　　　　　　比…厚

curved artfully but the curved lines are different on the top from
彎曲成…　巧妙的　　　　　　　　　　　　　　　　　上面

on the bottom. ⑨The lifting force is produced more efficiently by
　下面　　　　　　　　　　　　　　　　　　　　　更有效率

this artful shape of the wings. (see p.205)
　　　　形狀

問78　飛行機はなぜ、飛ぶことができるの？

答 ①下敷きのような**プラスチックの板**を、前方が上になるように傾けて持ち、その
墊板

まま走ると、下敷きが上にあがろうとする力を感じるでしょう。②**飛行機**が飛ぶのも
飛機

同じような**しくみ**です。

③飛行機が**前に進む**と、**翼**が風を受けます。④翼を通り過ぎた風は、少しだけ**下方に向**
機翼　　　　　　　　　　　　　　　　　　　　　　　　　　　　　　下方

きを変えます。⑤その結果、翼には上向きの力が加わります。⑥**下向きの空気の流れ**の
結果　　　　　　　　　　　　　　　　　　　　　　　　　　　　　空氣

反作用として生まれる力です。⑦これが、飛行機の**機体を持ち上げる揚力**となりま
反作用力　　　　　　　　　　　　　　　　機體　　　　　揚力

す。

⑧現在の**ジェット旅客機**などの翼は厚みがあり、**上面**と**下面**が異なる複雑な**カーブ**を
旅客機　　　　　　　　　　　　　　　　　上面　　下面　　　　複雑

描いています。⑨この複雑な**形**によって、**より効率よく揚力**が生まれるのです。
効率　　揚力

問78　為什麼飛機可以飛？

答 ①拿著一塊像墊板一樣的**塑膠板**，以前方朝上的方式傾斜，維持著這個姿勢跑步，是不是
會感覺到塑膠板有向上漂浮的力量呢？②**飛機**飛行也是運用同樣的**機制**。
③飛機在**前進**的時候，**機翼會迎面**受風。④通過機翼的風會些微地朝**下方改變方向**。⑤其結果會
施加向上的力量給機翼。⑥這是藉由**向下空氣流動**的**反作用力**所產生的力量。⑦這就形成了**抬起**
飛機機體的揚力。
⑧現在的**噴射旅客機**等機翼都很厚，上面與下面形成了不一樣的複雜曲線。⑨藉由這個複雜的**形**
狀，可以產生**更有效率**的揚力。

You can talk on the cell-phone anywhere within the area of the cell-phone network because the cell-phone switching station knows where every cell-phone is.

どこにいても携帯電話がつながるのは、交換局が
すべての電話機の位置を把握しているから

不論你在任何地方都能藉由行動電話來聯繫的原因，
是因為行動電話交換中心掌握了所有電話機的位置

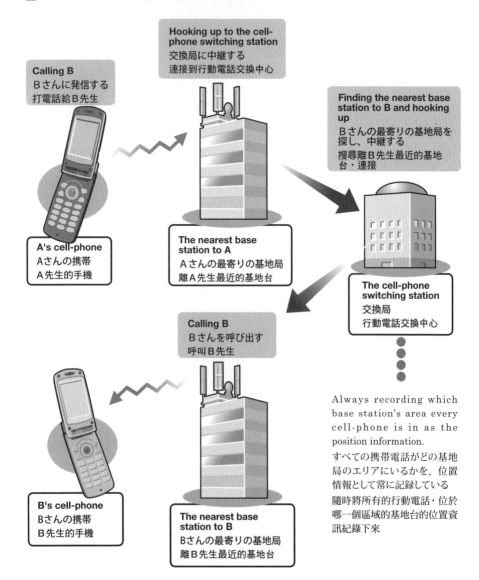

Hooking up to the cell-phone switching station
交換局に中継する
連接到行動電話交換中心

Calling B
Bさんに発信する
打電話給B先生

Finding the nearest base station to B and hooking up
Bさんの最寄りの基地局を探し、中継する
搜尋離B先生最近的基地台，連接

A's cell-phone
Aさんの携帯
A先生的手機

The nearest base station to A
Aさんの最寄りの基地局
離A先生最近的基地台

The cell-phone switching station
交換局
行動電話交換中心

Calling B
Bさんを呼び出す
呼叫B先生

Always recording which base station's area every cell-phone is in as the position information.

すべての携帯電話がどの基地局のエリアにいるかを、位置情報として常に記録している

隨時將所有的行動電話，位於哪一個區域的基地台的位置資訊紀錄下來

B's cell-phone
Bさんの携帯
B先生的手機

The nearest base station to B
Bさんの最寄りの基地局
離B先生最近的基地台

Airplanes can fly by changing flow of air to lifting power.

飛行機は空気の流れを揚力に変えて飛ぶ
飛機藉由空氣流動改變揚力而飛行

When you hold a plastic board with sloping upwards to its front end and you run as fast as possible, you feel the power of the plastic board to try to move upwards.

プラスチックの下敷きを、進行方向が少し上になるように傾けて持ち、全速力で走ると下敷きが上にあがろうとする力を感じる

當你以行進方向稍微傾斜向上的方式拿著塑膠墊板，以全力跑步的時候，墊板就會感受到向上的力量。

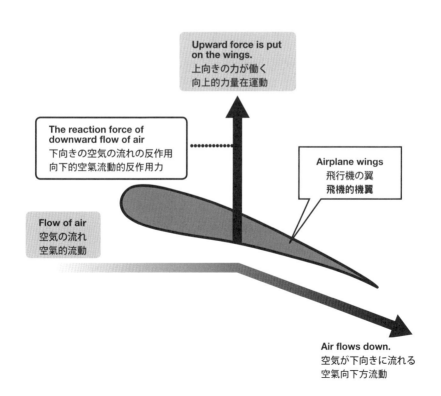

Upward force is put on the wings.
上向きの力が働く
向上的力量在運動

The reaction force of downward flow of air
下向きの空気の流れの反作用
向下的空氣流動的反作用力

Airplane wings
飛行機の翼
飛機的機翼

Flow of air
空気の流れ
空氣的流動

Air flows down.
空気が下向きに流れる
空氣向下方流動

Q79 Is it possible to make a time machine?

①It is thought in theory that you can travel from the present into the future. ②Einstein's special theory of relativity is the basis for this and it says that time stops when it moves at the speed of light. ③This means that time would stop in a rocket traveling at close to the speed of light. ④Imagine that a 30-year-old astronaut travels to and from a star about 10 light years away in a rocket at close to the speed of light in 20 years. ⑤The astronaut comes back to the earth 20 years later and his classmates in school are 50 years old at this time but the astronaut is still 30 years old. ⑥This time delay at high speed was already proved by the fact that clocks run slow in a jet fighter. ⑦It may be possible to make a time machine to travel into the future by using this fact.

⑧There are various ideas about a time machine which would travel from the present into the past. ⑨However, today, most scientists think making it is impossible.

問79　タイムマシンはつくれるの？

答 ①**現在**から**未来**への移動は、**理論的には**可能だと考えられています。②その根拠
となる**アインシュタイン**の『**特殊相対性理論**』では、「光速になると時間は静止する」とされています。③**光速に近い速度**で飛んでいる**ロケット**があるとすると、その中の時間は止まっているという意味です。④仮に30歳の**宇宙飛行士**が10光年離れた星を、**20年**かけて光速に近い速度のロケットで**往復する**とします。⑤宇宙飛行士が地球に戻るのは20年後であり、**同級生**は50歳になっているのに、宇宙飛行士は30歳のままということになります。⑥こうした高速移動時の**時間の遅れ**は、ジェット戦闘機の中の時計が遅れることなどで、すでに証明されています。⑦これを利用すれば、**未来へ行くタイムマシン**が実現する**可能性があります**。

⑧現在から**過去**へ行くタイムマシンについても、さまざまな仮説が立てられています。⑨しかし現在のところ、これは不可能だというのが多数の意見です。

問79　時光機器有可能製造得出來嗎？

答 ①由現在移動到未來，理論上是有可能的。②根據愛因斯坦的『**特殊相對論**』所說，「當以光速移動的時候，時間是靜止的。」③也就是說，若有一架**火箭**以**接近光速的速度**飛行時，當中的時間是靜止的意思。④假設有一位30歲的**太空人**，搭乘火箭花了**20年**時間，以接近光速的速度**往返**距離10光年以外的星星。⑤當20年後**太空人**返回地球，雖然**同班同學**的人已經50歲了，但是太空人依舊是維持著30歲的樣子。⑥像這樣高速移動時的**時間延遲**，在噴射戰鬥機中的時鐘延遲等案例中，已經得到了**證實**。⑦假如利用這個原理，想**前往未來的時光機器**就有實現的**可能性**。
⑧關於由現在前往**過去**的時光機器，有著各式各樣的假說。⑨但是到目前為止，這是不可能的意見佔大多數。

TITLE

中英日對譯 圖解科學Q&A

STAFF

出版	三悅文化圖書事業有限公司
監修	松森靖夫
英文監譯	古家貴雄
譯者	闕韻哲

總編輯	郭湘齡
責任編輯	王瓊苹
文字編輯	林修敏　黃雅琳
美術編輯	李宜靜
排版	執筆者設計工作室
製版	明宏彩色照相製版股份有限公司
印刷	綋億彩色印刷股份有限公司
法律顧問	經兆國際法律事務所　黃沛聲律師

代理發行	瑞昇文化事業股份有限公司
地址	新北市中和區景平路464巷2弄1-4號
電話	(02)2945-3191
傳真	(02)2945-3190
網址	www.rising-books.com.tw
e-Mail	resing@ms34.hinet.net

劃撥帳號	19598343
戶名	瑞昇文化事業股份有限公司

初版日期	2012年1月
定價	250元

ORIGINAL JAPANESE EDITION STAFF

裝幀	杉本欣右
イラスト	笹森　識
日本文執筆	中村英良＋森井美紀
英文執筆	中堂良紀
DTP・編集	スタジオスパーク

國家圖書館出版品預行編目資料

中英日對譯 圖解科學Q&A／
松森靖夫監修；闕韻哲譯.
-- 初版. -- 新北市：三悅文化圖書，2011.12
208面；14.8×21公分

ISBN 978-986-6180-85-9 (平裝)

1.科學　2.問題集　3.通俗作品

302.2　　　　　　　　　　100026352

SOBOKU NA "?" GA YOKU WAKARU! EIGO TAIYAKU DE YOMU KAGAKU NO GIMON
(Questions and Answers about Science in Simple English)
supervised by Yasuo Matsumori, Takao Furuya
Copyright © 2010 studiospark+A Priori LLC
All rights reserved.
Originally published in Japan by JITSUGYO NO NIHON SHA, Tokyo.
Chinese (in complex character only) translation rights arranged with
JITSUGYO NO NIHON SHA, Japan
through THE SAKAI AGENCY and HONGZU ENTERPRISE CO., LTD..